IN THE TEMPLE
OF THE RAIN GOD

To: MADELINE COTHER

WHO WAS THERE

Garrett Wilson

IN THE TEMPLE
OF THE RAIN GOD:

The Life and Times of "Irish" Charlie Wilson

GARRETT WILSON

University
of Regina   CPRC
               PRESS

Printed and bound in Canada at Marquis Imprimeur inc.
The text of this book is printed on 100% post-consumer recycled paper with earth-friendly vegetable-based inks.

Cover and text design: Duncan Campbell, CPRC. Editor for the Press: Brian Mlazgar, CPRC.

Library and Archives Canada Cataloguing in Publication

Wilson, garrett, 1932–
In the temple of the rain god : the life and times of "irish" Charlie wilson / garrett wilson.

(Trade books based in scholarship, ISSN 1482-9886 ; 35)
Includes bibliographical references.
ISBN 978-0-88977-288-5

1. Wilson, charlie, 1884-1970. 2. Irish canadians— Saskatchewan—biography. 3. Immigrants—saskatchewan— Biography. 4. Farmers—saskatchewan—biography. 5. Businesspeople—saskatchewan—biography. 6. Farmers—Saskatchewan—economic conditions—20th century. 7. Agriculture—saskatchewan—finance—history—20th century. 8. Agriculture—saskatchewan—history—20th century. 9. Saskatchewan—economic conditions—1905-. 10. Saskatchewan—history—1905-. 11. Saskatchewan—Biography. I. Title. Ii. Series: TBS ; 35

Fc3522.1.W58w34 2012   971.24'02092   C2012-901985-2

10   9   8   7   6   5   4   3   2   1

CPRC 𐂀
P R E S S

Canadian Plains Research Center Press, University of Regina
Regina, Saskatchewan, Canada, S4S 0A2
TEL: (306) 585-4758 FAX: (306) 585-4699
E-MAIL: canadian.plains@uregina.ca WEB: www.cprcpress.ca

We acknowledge the financial support of the Government of Canada through the Canada Book Fund for our publishing activities, and the Creative Industry Growth and Sustainability program which is made possible through funding provided to the Saskatchewan Arts Board by the Government of Saskatchewan through the Ministry of Tourism, Parks, Culture and Sport.

 Canadian Heritage   Patrimoine canadien

 Government of Saskatchewan

 SASKATCHEWAN ARTS BOARD

In Memoriam to
Kevin D. Wilson,
who was much of the story

Moira Vine,
who first imagined this book

And to
Sheila McMullan,
for love and encouragement

CONTENTS

Foreword—ix
Preface—xii

Chapter 1:   WICKLOW TO LONDON 1

Chapter 2:   TO CANADA 10

Chapter 3:   GRAIN BUYER 15

Chapter 4:   HOMESTEADER 25

Chapter 5:   TO THE SOUTH COUNTRY 35

Chapter 6:   WILSON BROTHERS and VICTORIA TRUST 45

Chapter 7:   1918 FLU EPIDEMIC 55

Chapter 8:   POST-WAR PROSPERITY and DISTRESS.......................................63

Chapter 9:   WILSON BROTHERS DISSOLVE 73

Chapter 10: PROSPEROUS DAYS 81

Chapter 11: DARK DAYS ARRIVE 89

Chapter 12: DROUGHT 99

Chapter 13: THE LIMERICK PLAN 107

Chapter 14: THE FARMERS' CREDITORS ARRANGEMENT ACT 116

Chapter 15: DEBT ADJUSTMENT 127

Chapter 16: 1936 145

Chapter 17: THE BOARD OF REVIEW 155

Chapter 18: 1937: THE YEAR FROM HELL 165

Chapter 19: 1938 175

Chapter 20: 1939: RAIN RETURNS 183

Chapter 21: AN ADVENTURE IN WHEAT 192

Chapter 22: THE WAR ENDS and PROSPERITY RETURNS............................203

End Note—212
Epilogue—214
APPENDIX A: "The Limerick Plan"—219
APPENDIX B: Biographies—225
Bibliography—229

FOREWORD

In the 1960s, when my older sister Moira was teaching school in Burns Lake in northern British Columbia, she spent some part of her summers in Saskatchewan at our family home in Limerick. There she embarked on a project to record an account of our father's career. A tape recorder was provided to allow Charles Wilson to dictate his recollections, which were then transcribed. Charles died in 1970, and the project seemed to stall. When Moira died at Burns Lake in 1984, I retrieved the suitcase in which she had deposited the transcripts, along with her notes and files, and brought it home, not knowing much about it all.

The suitcase remained undisturbed until just a few years ago, when I finally looked through the contents. I found far more material, and history, than I had known existed. I was the youngest of our family of four children, thirteen years younger than Moira and nearly half a century younger than our father. Because of that gap of time, I had been unaware of much of his career. Moira had been much closer to the scene, had been acquainted with many of our father's contemporaries, and had known which documents to preserve. In that suitcase lay the story of an Irishman's journey through the first fifty years of Saskatchewan's history, from farm and construction labourer to homesteader to business success and then through Depression insolvency to the farm debt crisis and the Second World War. Much of the story was a revelation to me, but there was a book in the suitcase, and I perceived an obligation to complete the work.

The tapes of Charles' dictated recollections were with the files. I doubted that the flimsy cassettes had survived forty years of careless storage, but Rolf Winkler, an audiovisual wizard, was able to transfer the words to digital compact discs of broadcast quality. To ensure that nothing had been omitted from the original transcripts, old friend Lucy Balfour again transcribed the entire record.

My previous writings had given me a fair knowledge of the history of Saskatchewan's early years, particularly those known as the Dirty Thirties, and I eagerly set to the research needed to flesh out that knowledge.

Much of the needed history lay with the Saskatchewan legislature. Unfortunately, since there was no Hansard record of proceedings until 1947, the political debates of the 1920s and 1930s were not preserved and can be gleaned only from contemporary newspaper accounts, a sad omission. However, the legislature's journals contain some of the more significant speeches of government ministers, enough to impel a researcher to conclude that the quality of debate in the legislature has deteriorated significantly over the years. The staff at the Legislative Library were gracious and helpful.

Charles spent seven years, from 1936 to 1943, with the Saskatchewan Board of Review created by the Farmers' Creditors Arrangement Act of 1934. Saskatchewan Archives and Library and Archives Canada held next to nothing on this important feature of the struggle to deal with the farm debt crisis on the western prairies during the 1930s. Fortunately Moira had preserved a substantial documentary record of the accomplishments of the board in Saskatchewan.

She had also saved many of the annual reports from the 1920s and 1930s of Victoria Trust and Savings Company of Lindsay, Ontario, with which Charles had had a fifty-year association. Those reports, providing an Ontario lender's view of the success, agony, and recovery of the Saskatchewan farm sector, now seem to exist nowhere else. Victoria Trust has long since been absorbed by ScotiaBank.

Beginning about 1928, Charles maintained a skeleton diary, consisting mostly of weather notes, crop conditions, and some mention of his travels but, unfortunately, very little opinion. These diaries, or copies, have been preserved, again providing a window onto Saskatchewan's past.

Also retained were leather-bound ledgers containing the financial history of Wilson Brothers, from its beginning in 1912 straight through the following fifty years. These ledgers have been deposited with Saskatchewan Archives, as will be the rest of the material, including a fair collection of correspondence, following publication of this book.

Rob Nestor again provided excellent research, digging out papers and articles from the 1930s dealing with the economic and financial sides of the farm debt crisis.

Most of our carefully preserved family photographic archives have disappeared, but Gerald Matthews, with his lifetime of photographic experience, was able to restore several of the remaining items. Saskatchewan Archives was able to supply others illustrating the physical damage to the land in the Dirty Thirties. Tim Novak has a remarkable familiarity with the Archives' photo collection.

Professor Bill Waiser of the University of Saskatchewan was again very helpful with advice and in referring me to invaluable sources.

Brian Mlazgar, publications manager of the Canadian Plains Research Center at the University of Regina, as usual performed magnificently, spotting merit in the early drafts, providing guidance and advice, and turning the whole into an interesting historical record. I am profoundly grateful to Brian for his expertise and kindness.

The result of all this is a portrayal of social, agricultural, political, and economic life in Saskatchewan during its first fifty years, much of it through the eyes of one Irish immigrant.

Garrett Wilson
Regina, Saskatchewan
March 2012

PREFACE

Roll forth, my song, like the rushing river,
That sweeps along to the mighty sea;
God will inspire me, while I deliver
My soul of thee.[1]

That, at least, is my earnest hope and my inspiration in starting
this little volume.

It is not intended to be either history or romance. I hope it may
prove to be an accurate, down-to-earth record of the settlement and the
development of that large area of Saskatchewan south of the mainline of
the Canadian Pacific Railway which has come to be known as the South
Country. From the first moment I located in it, I was impressed by the
quality of the settlers who chose the South Country as a place to establish
a home, and I still think that no part of Canada was more fortunate
in the quality and calibre of the people who came in and occupied these
lands. They came from everywhere in Canada, from Jordan and Bruce
Counties; from Joliette, Quebec; from Lunenburg, Nova Scotia; from
Newfoundland and everywhere else. They must have come out of fine
Christian homes, because they brought with them the instincts of good
citizenship and community partnership.

They came into a country that was absolutely raw—as it came from the
hands of God—and had none of the requisites of civilized living within
it. There were no schools, no churches, no rural municipalities, no roads,
and these are the very fundamentals of life in a new country. There was,

however, without limit, a community spirit which enabled the early settlers to work together and establish the rudiments of the first civilization here.

When I say they came into an absolutely virgin country, I mean exactly that. I can imagine that, when my own father-in-law, Davis Sproule, settled on the E 1/2 15-7-5 W3rd, his first concern was to erect a house of some kind, in which he could shelter the family he had brought with him from Nova Scotia from the natural elements. He went about this job with the aid of logs from Wood Mountain and sods from his homestead, and with these simple materials he erected a large home which was warm and comfortable, even if it was as "homely" as any building could be.

And I can imagine that Mr. Sproule, when he got the building erected and had time to straighten his back in the yard, looked north and saw his neighbour moving around on his quarter section, and possibly a thought came to him: "I notice that there are quite a number of children in this community already, and they should have facilities for attending school at the earliest possible date. I see my neighbour over there. I'll go over and have a talk with him."

He did that, and the two men conferred—across the fence, if a fence existed then—and decided that a school was a necessity in the community. Possibly they went to Regina and got details of the simple steps which had to be taken to establish such a school. No doubt a few days later they held a larger meeting, and it was decided to build a modest school. Six months later, probably, the school was there. This provided a very convenient place for church services, and, in the minds of these early settlers, it was just as important to have church services in the community as to break up the land and try to grow wheat on it.

Probably, a little later on, they got discussing a couple of long pieces of road which were low lying and impassable in wet weather—although the prairie trails, which were sound, provided pretty acceptable roads for the rest of the picture. Another meeting was no doubt called, and the initial steps were taken to form municipalities.

And now the country was well on the way to civilization, entirely through local initiative, not waiting for the government at Regina to send organizers in to do for them the things which they knew perfectly well they were quite capable of doing for themselves.

I have lived amongst them a long time, and I remember still being impressed, from my earliest association with them, by the excellent amount of good horse sense and ability which, as a group, they possessed. They built according to their means, one step at a time, and they built well.

It is my hope, if God gives me the health to persist with this volume, to make it to some degree a record of many of these men who made deep, and important, and enduring footprints in the sands of time in the South Country. There was at all times amongst them a community spirit and a spirit of neighbourliness which made every home a stopping place for the weary traveller, where he could be sure at least of a warm welcome.

In my early days, driving around these prairies with a team and buggy, when the team got tired in the afternoon, I was often too far from town to think of driving them there to stay overnight. I simply headed for the nearest little house which was showing smoke and requested sanctuary there. I cannot recall a single incident when I was denied a welcome and an invitation. When it came to bedtime, it might be necessary for me to go to my buggy and bring in a couple of horse blankets for the purpose of making my own bed on the floor, the supplies in the house being barely adequate to the little family which was already there. I spent many pleasant and happy evenings in those homes, and I carry with me in my heart a gratitude to those men and women which can only die with me.

I confided to my old friend, George F. Edwards, formerly of Markinch and now of Vancouver, some years ago the idea that I might undertake this volume. He gave me very cordial support in the idea and then said he would furnish me with a title. The title proved to be "They Were Giants in Those Days."

I rejected it out of hand and told him, "George, that title implies that we have run out of giants, that there are none in these days. But I happen to have enough contacts with the young people of the South Country to convince me that their ranks contain many specimens who are quite capable of developing into giants, given the opportunity or the provocation. So I will not use that title." I said, "I have thought of one of my own, and I think you will agree with it. I propose to call the volume In the Temple of the Rain God because I have noticed in Saskatchewan that, amongst the many other things which survived us, there is one thing to which we all render abject allegiance, and that is our veneration for rainfall."

Indeed, I have intended for some years to use some of the nice cobblestones which are easy to get hold of here and build in the northwest corner of the garden a shrine to the Rain God, topped by a small fountain. I have never gotten around to this, and probably now I never will. But rain, and the respect and veneration for it, are the only things on which I think anyone could say the people of Saskatchewan are absolutely united.

Finally, by way of completing my preface to this little volume, I wish to dedicate it to the original and the present-day settlers of the South Country. A great many of the oldtimers have passed on, and the country is more thinly settled by far than it used to be, but I think I can still assure the traveller that, if he ever gets caught in a storm and approaches one of our farm homes, he will find sanctuary, and a welcome, and a great deal of comfort besides. In that respect, I do not think that the settlers have undergone any change.

I did not myself homestead in the South Country, but rather up in the Humboldt area, west of Dana, where my brother and I secured filings on the S 1/2 22-38-26 W2nd. We had quite an experience up there, and I cannot exclude it, because the money I received from the sale of my homestead enabled me to make tracks for the South Country, which I loved so well at first sight that I have never sought another homesite since.

I am going to include quite a lengthy account of my personal experience on the homestead. Then I will come back to the South Country and stay there, as I have done in my lifetime, and endeavour to make a record of all those fine and valuable citizens whom it has been my privilege to know since I settled at Limerick.

I have said the book is not a romance, but I do not intend to exclude from it the story of my meeting with the greatest lady, who, by God's grace, still shares my home with me. That would be an unforgivable omission because, for a period of well over fifty years now, she has been my trusted and invaluable partner in every section of my life, and, when I thank God for the many blessings which I enjoy at this moment, she is the first item that I render him gratitude for.

I am fully aware that I have put off for too long the attempt to write this book. I am now eighty-six years of age, and I warn my readers, if I have any, that my memory, although it was a most remarkable memory throughout my life, cannot be expected to be entirely reliable at the advanced age I have achieved. However, I will endeavour to avoid, as far as an Irishman can, indulging in exaggeration, and on the whole I hope to produce a simple record of life in this province, and particularly in the South Country, which will be reasonably accurate.

With that I terminate my preface to this book.

ENDNOTES

1. James Clarence Mangan, "The Nameless One."

CHAPTER 1

WICKLOW TO LONDON

Charles Wilson, a young Irishman alone in London, England, spent a sleepless night, July 1, 1905, wrestling with a life-altering decision. The mail that day had brought a package from his friend Ben Lloyd in Dublin—a ticket to Canada. Ben had booked passage for them on the *Virginian*, sailing from Liverpool on July 5th, only four days away. Destination Quebec, Canada.

True, Charles had told Ben that he would go with him to Canada that year. The two of them had decided more than a year earlier that they would emigrate and stake their futures in one of the faraway colonies. They had pored over the immigration literature from Canada, New Zealand, and Australia. Ben, keen on Australia, strutted about London in what he hoped were Australian clothes: a rough red shirt, corduroy breeches well stained with patches of pitch, and heavy, hob-nailed boots. Both were taken by an evocative poem that struck a strong chord with the country-boy Irishmen stuck in the centre of London, then the largest city in the world, with a population of 7 million.

Clancy of the Overflow
I had written him a letter which I had, for want of better
knowledge, sent to where I met him down the Lachlan, years ago.

He was shearing when I knew him, so I sent the letter to him,
just on spec, addressed as follows, "Clancy of The Overflow."
And an answer came directed in a writing unexpected,
(And I think the same was written with a thumb-nail dipped in tar)
Twas his shearing mate who wrote it, and verbatim I will
    quote it:
"Clancy's gone to Queensland droving, and we don't know
    where he are."
In my wild erratic fancy visions come to me of Clancy
gone a-droving down the Cooper where the Western drovers go;
As the stock are slowly stringing, Clancy rides behind them
    singing,
for the drover's life has pleasures that the townsfolk never know.
And the bush hath friends to meet him, and their kindly voices
    greet him
in the murmur of the breezes and the river on its bars,
and he sees the vision splendid of the sunlit plains extended,
and at night the wond'rous glory of the everlasting stars.
I am sitting in my dingy little office, where a stingy
ray of sunlight feebly down between the houses tall,
and the fetid air and gritty of the dusty, dirty city
through the open window floating, spreads its foulness over all.
And in place of lowing cattle, I can hear the fiendish rattle
of the tramways and the buses making hurry down the street,
and the language uninviting of the gutter children fighting,
comes fitfully and faintly through the ceaseless tramp of feet.
And the hurrying people daunt me, and their pallid faces
    haunt me
as they shoulder one another in the rush and nervous haste,
with their eager eyes and greedy, and their stunted forms and weedy,
for townsfolk have no time to grow, they have no time to waste.
And I somehow rather fancy that I'd like to change with Clancy,
like to take a turn at droving where the seasons come and go,
while he faced the round eternal of the cash-book and the
    journal —
But I doubt he'd suit the office, Clancy of The Overflow. [1]

But Clancy and Australia had lost out. Canada it would be.

I was having quite an enjoyable time in London in the Customs Service. We had a big sports club, known as the Customs Sports Club, and I was general secretary of it. Included in our activities were a rowing club and a rugby football club, to both of which I belonged actively so that I spent most of my time on the river Thames in the summer and on the football field in the winter. I was enjoying life immensely and pretty well forgot about the promise I had made to Ben. Ben never forgot it, and he went to the office of the Allan Line and booked two tickets for Canada, sending them to me because he like me did not have the funds to pay for them.

I at the time was very busy on the river. We had an eight-oared boat entered in the Grand Challenge Cup at Henley, and the keenest ambition I think that I ever had in my life was to row at Henley. Even if you came in last, it was a distinction to have been in the Grand Challenge Cup.

I did have the price of a ticket to Canada. When I was born, my parents named me Charles, thinking that I would stand a good chance of finding a place in the will of my grandfather of the same name. Their guess was correct. My grandfather altered his will to the extent of leaving me £100 to be paid to me on my twenty-first birthday. Had it not been for that little circumstance, I doubt that I ever would have been able to accumulate enough money to buy a ticket to far-off Canada. I entered the Customs Service at a salary of £70 a year, which had to cover all my living, board, and clothes and recreation, together with the annual trip to Ireland, which I considered a must. There was nothing left, of course, even in those days of extremely low prices when I could buy an excellent suit of clothes for the sum of two guineas.

During those years, I had forgotten all about being included in his will, but on March 21, 1905, my twenty-first birthday, I received notice from the solicitors that they had £100 to my credit. I immediately instructed them to send it along, and it came in the form of an investment which was very common in those days, British Consols. I liquidated it in short order.

I had noticed that the first-class clerk in the office where I worked had a bank account and a cheque book, which seemed to me to be a very dignified way of doing business. Accordingly the next day I walked into

William Deaton's bank on Lombard Street with my £100, and when I got the attention of a teller I informed him that I wished to open a bank account and get a cheque book.

The gentleman was very courteous but also very shocked and surprised. He asked me to wait a couple of minutes and then went into one of the senior offices and conferred with the gentleman there. Presently they took me in there, where I found a very courteous and dignified gentleman wearing a claw hammer coat and striped pants who sat me down and explained to me in detail why it was impossible for them to accept money casually like that. He told me that it would be necessary for me to obtain an introduction to the bank. But he made it clear enough, also, that they were not interested to any great extent in such a little gob as I had to offer.

I left the bank five minutes later feeling about the same as I would have felt if he had put me over his knee and spanked me. I walked down the street a couple of blocks to where I knew the post office savings bank was located. There they accepted my little deposit with alacrity, and there it stayed until I used up the final balance of it to pay for my homestead shack on the SW 1/4 22-38-26.

On the first of July, 1905, I received a letter from Ben Lloyd enclosing the two tickets to Canada with a deposit paid on them and a request to me to take up the balance. Here was where the £100 I had received from my grandfather came into real use.

I went to bed that night most disturbed. The boat was due to leave Liverpool on the 5th of July. The Henley Royal Regatta was scheduled for July 11th. I was rowing bow oar, and to leave on such short notice was most unfair to my colleagues in the rowing crew.

On the other hand, when I thought the problem through, I remembered that every fall I had promised Ben Lloyd that I would break off and come in the spring. In the spring, I had pleaded for six months more in order to enjoy another season on the river, and I knew instinctively that the soft life was getting to me, and it would be like this forever. Accordingly I woke up in the morning with my mind made up that I would come now.

I went to my senior clerk and told him that I wished to resign, that I proposed to sail for Canada July 5th. He was shocked and dumbfounded and told me, "Mr. Wilson, you cannot do that. You are bound to give a month's notice."

I thought for a moment and then said to him, "Mr. Stevenson, I am giving you a month's notice now with a warning that I am sailing for Canada on the 5ᵗʰ."

The Customs people were very nice about it. When they decided I had made up my mind, they cooperated with me in every way. I went in one day to say good-bye to the secretary of the Board of Commissioners, one A.J. Dite. Mr. Dite said something to me which could easily have been my ruin the next two or three months. He told me, "Charlie, if this enterprise fails to be the El Dorado that you are expecting, before two or three months have elapsed, apply for reinstatement. You will be sure to get it."

His rural roots were reclaiming the Irish country boy—but not back to Ireland. Instead, Charles would head to a new life on the virgin lands of western Canada.

I was born in the townland of Driem, near the crossroads village of Rathdangan, in the Barony of Upper Talbotstown and County of Wicklow, in Ireland. It was near the home of Charles Stewart Parnell, whose life was not without influence on mine.

I was born on March 21, 1884, the son of William and Sarah Wilson, nee Willoughby. My parents farmed eighty-two acres as a stock and dairy farm on the high land between two rivers, both tributaries of the Slaney, hence the name of the townland, the Gaelic word Drim, or Driem, a back.

The farm was held by William Wilson as tenant, as most of Ireland's farms were at the time. Title lay with large landowners, usually of English origin, until Irish Land Reform. Sixteen milking cows provided the annual cash rent of thirty-five pounds. The small Wilson clan in County Wicklow was part of a Protestant enclave in mostly Catholic southern Ireland. They shared the widespread and fierce nationalistic sentiment that simmered with latent hostility toward the English, who had ruled Ireland since the Tudor conquest in 1603. Not until 1922, after years of violent guerrilla warfare, did the southern twenty counties achieve independence, leaving the six counties of Northern Ireland in England's grasp. Land Reform enabled William, Charles' younger brother, who had succeeded to Driem, to acquire title to the small farm. William had undertaken the customary arrangement of caring for his parents during their remaining years.

I went to the National School at Rathdangan, under the teachership of one Richard Griffin, to whom I wish now to record a belated gratitude. He was a great and conscientious teacher. If I have found myself possessed through life of an education adequate to the responsibilities which have fallen to my lot, I owe it very largely to Richard Griffin.

My school days were very simple. We had no gym or YMCAs *and no conception that it was up to the older people to do something for us in the way of providing us with entertainment. We had the fields and the mountains [the Wicklow Mountains, very modest by Canadian standards] and the very lovely little rivers which emerged from the mountains, and we seemed to have no difficulty at all in finding our own amusement amongst those surroundings.*

Life was very strict. There was right, and there was wrong, and you were taught what both of them meant. There was no such thing as being half right or half wrong. If it wasn't right, it did not qualify, and if it was wrong, it was out.

I belong to the first generation that went to school. Previous to my time, no government in Ireland, or, I think, in Europe, had admitted that education was a public responsibility. When the Irish government started to build schools in the rural sections of Ireland, they were faced with the difficulty that they had no backlog of teachers to draw from. There had been no teachers' colleges, and in many cases they were obliged after building their school to start it up with some resident of the district who had a little knowledge. I call to mind the case of Mine School, where, apparently, no one was available except a farmer, a certain Mr. Byron. I knew Mr. Byron, and it is very certain that he had never been to school. Yet he held down Mine School for a period of about sixty years. When my brother Tom returned from World War One, he informed [me] that Mr. Byron had just retired. I expressed amazement and said, "He was never fit to be in a school, even the first day it opened. How did it come about?"

He replied without hesitation. "Your friend Canon Willis, the Protestant vicar, had the whole say about it, and it was not the canon's idea that his people should absorb too much education. He preferred that his hillbilly parishioners, when they met him on the road, should continue the practice of removing their hats before passing him, and so Mr. Byron suited him right down to the ground."

Looking back over the long years, it seems to me that I enjoyed my schooldays very much and did not have a dreary or a boring time at any stage in the year. In addition to the society of my father and mother, I had a warm personal relationship with an old Irish friend, a neighbour. He would occasionally come over in the morning and pick me up and take me for a day's rabbit shooting, because he was a great hunter. He did that in 1897, on the eve of my departure for a boarding school where I had secured a scholarship. I remember still that afternoon, when my friend John Bourne was saying good-bye to me, he remarked that probably my parents had taught me everything I needed to know on the eve of leaving home. "But," he said, "I will take a chance on saying one thing, which I hope you will remember. Keep away from bad company in life, and you will never find yourself in jail." I have never forgotten his behest, and I am happy to say I have never been in jail.

Having reached the limits of his school, he [Richard Griffin] prepared me for a competitive examination for a scholarship at Farra School, near Mullingar, Westmeath County, one of a series of schools maintained by the Society for the Promotion of Protestantism in Ireland, and to this society I owe also a deep debt of gratitude.

I won the scholarship at Farra and went there in August 1897 for a two-year course, under the headmastership of Thomas C. Foster, MA. For all that he was a harsh and stern man, he was my good friend. I might as well say here that, on looking back on my life, I have had more good and loyal friends than I have ever deserved, commencing with Mr. Foster or maybe before. In my second year, we had a mathematical teacher, one F.W. Allen (Turfy Allen), to whom as a teacher I think I owe the greatest debt of all, because he took all the drudgery out of the job of acquiring an education for me and first opened my eyes to what it was all about and where it was leading.

It was usual on the expiration of the two-year scholarship for the head to recommend two or three boys for a third year. Mr. Foster had generously recommended me for the additional year, and I was sent to the Educational Institute at Dundalk for that, August 1899–June 1900.

At the end of that year, I was successful in securing a two-year scholarship at Santry, County Finegal, and arrived there in August 1900. The head was a Mr. McLelland, a harsh and stern man for whom I entertained no love or respect. All the boys at Santry were seventeen or over. In February 1901, two of them were reported to the head for cigarette smoking. He flogged them in front of the assembled school, cruelly and unmercifully. That made up my mind to abandon my scholarship. I used to smoke an odd cigarette too. I did not intend to accept the mauling those two boys got, and I could figure no other way of being safe except to quit the school. Apparently the idea of not smoking never occurred to me, an interesting comment on prohibition.

A week or so later I went to Dublin with the football team to play against Harcourt Street High School, and I did not return to Santry.

Instead, I went to Skerry's Civil Service College, Stephen's Green, Dublin, and in October 1901 took at the same sessions the examination for the Second Division of the Civil Service and Customs, Outport Clerkships, Indoor Service, selecting the latter in case of success in both. I was successful in both and in Customs ranked first in Ireland and second in the United Kingdom.

In January 1902, I was assigned to the Custom House in London and entered on my duties.

Life in London was very enjoyable. Plenty of opportunity for rugby football, rowing, swimming, and all other sports. I was general secretary of the Custom Sports Club and had won the coveted blue velvet and gold braid cap for football, a medal for rowing, another for the mile swim. But always the peasant boy in me was in rebellion against life in a city. The stone streets burned my feet, and my every instinct turned to green fields and country roads.

While employed at the Custom House in London, Charles had returned each year to his family home. The sudden departure in July 1905 left no time for even notice to his parents, much less a last visit, and he never saw them again. It was just one year short of half a century later when he next visited Ireland.

ENDNOTES

1. Andrew Barton, "Clancy of the Overflow." Barton (1864–1941) was known as Banjo Paterson, a popular Australian folk poet who also composed "Waltzing Matilda." At the age of eighty, Charles accurately quoted the total poem.

CHAPTER 2

TO CANADA

I made the voyage from Liverpool to Quebec aboard the Allan liner Virginian. It was almost her maiden voyage, and she was a lovely boat, clean as a whistle when we left Liverpool.[1] She was, however, loaded to the gunnels with those settlers whom Clifford Sifton had encouraged to immigrate to Canada, the "stalwart peasants in sheepskin coats."[2]

The vast majority of them had never seen any of the equipments of civilization before they embarked on their trip. They were not even familiar with the functioning of a water closet, and, of course, three-quarters of them were violently seasick the first day out. What they did to that lovely new liner in the first twenty-four hours defies description. On the upper deck, they were lying on the floor in rows being sick, and there was nothing that could be done about it.

My friend Ben Lloyd and I were quartered in a cabin for four, and we had two Galacians for companions. We soon discovered that in the next cabin were two Englishmen and two more Galacians. I conceived the idea of moving the Galacians and replacing them with the Englishmen. The move was mutually satisfactory. The Galacians were anxious to rejoin their own people, and we were happy to have the two Englishmen.

The food in the common dining room was very indifferent. It had to be to stay within the very modest fare of $50.00 which the Allan line charged us from Liverpool to Quebec. When we got the two Englishmen

R.M.S. *Virginian*, (Allan Line) 12,000 Tons, Turbine.

The *Virginian*, aboard which Charles sailed to Canada in July 1905.

into our quarters, we were quite comfortable, and we took care of the food proposition by bribing the steward to bring us special food to the cabin. No doubt he was purloining it from the kitchen, but that was no concern of ours.

With this setup, we endured the nine days of the voyage very tolerably and arrived in Quebec in good shape. Upon stepping off the Virginian at the port of Quebec, I saw my first Canadian newspaper. The headline was "The Birth of the Two Provinces of Saskatchewan and Alberta," and I was headed for one of them.[3]

When we left the steamer at Quebec, we were taken over by the Canadian Immigration Department and shipped to Manitoba immediately. I took a look at the slat seats in the rail cars making up the special train in which it was proposed to ship us out, and I did not like them. I suggested to my companion that we would make a point of missing that train as we had tickets to Winnipeg, and the CPR was under obligation to take us there anyway.

That we did, and that evening we applied for passage on a regular CPR train, which had soft cushions instead of slat seats. The conductor, naturally, objected very strongly, but nonetheless he allowed us to board the train, and we got to Montreal in comfort. At Montreal, however, we overtook the immigrant train and were transferred back to it, and in it we made the rest of the trip to Winnipeg.

We were taken to Winnipeg in cars which would hardly be used as cattle cars nowadays. There was nothing you could break or steal. Hard wooden slats and nothing else. You had to make your own arrangements for sleeping. The railroad company did not furnish any food, and caught without warning the best we found it possible to do was lay in a supply of Clark's Pork and Beans and some loaves of bread, and on that we survived to Winnipeg. It was a most unpleasant and disagreeable voyage. There was no place to sleep except on the slats, and having no experience, of course, we had no blankets.

I forget how many days it required to make the trip, but we survived it, and we landed in Winnipeg still under the tutelage of the immigration people. We were dumped off for reshuffling in the big Immigration Hall there. We walked out of the station, and there were quite a number of ox teams in front of the station, lying down in their harness, chewing their cuds, and contributing a generous manure bank to the scenery. It was about 100 above in Winnipeg that day.

The immigration people were certainly efficient. We were in Winnipeg only a few hours when we were informed that we were allotted to Virden, Manitoba, and would leave for there the next morning since jobs were awaiting us there.

We complied and arrived in due course in Virden, where a couple of farmers were waiting for us. We were rather resentful of the manner in which we were being pushed around, and we did not altogether like the appearance of the two sharp-faced farmers who were waiting for us, so we turned their jobs down. Instead, we went to the hotel, the Wolseley Hotel across the street, rooms one dollar per day, spent one night in reasonable comfort, and the next morning, arranging with the hotel to store our luggage, we hit the road going south from Virden, feeling sure that we could rustle a couple of jobs on our own account.

We had not walked many miles before a farmer overtook us with a team and buggy and stopped to see if we were looking for jobs. The upshot of it was that we both hired out to him, Ben to work for one Edward Knight, I to work for Fred Reinhart and Jackson Meisner, who were *farming in partnership. The two farms were not far apart, about twelve miles south of Virden.*

I pay tribute to the efficiency of the Manitoba farmers of that time. By three o'clock in the afternoon, I was out in the field driving four horses on six sections of drag harrows in loose, dry summerfallow, walking behind them, no harrow guards, and I still remember that ordeal. The weather was hot, the fallow was dry, and you can imagine what I looked like at quitting time that evening, while my tongue was literally hanging out to my chin for want of a drink of water, which he had failed to provide. There were no facilities on the farm for taking a bath either. I was being paid the magnificent stipend of fifteen dollars a month, twenty-five dollars during harvest, and had a clear contract that I would stay until he finished threshing.

Crude surroundings. I had a cot to sleep on with a gunny sack filled with hay for a pillow, and I was got out of it every morning at five o'clock, and it would be nine or ten o'clock before I got a chance to get back to it too. Fifteen dollars a month. It was pretty considerable hardship, particularly since the farmer didn't try to help us in any respect. He should have insisted we get gloves right away. We didn't, and the mess that our hands became in a short time for the lack of working gloves was something cruel.

I had been accustomed to eating five times a day, including afternoon tea at four o'clock. Now I suffered from thirst, and I suffered from hunger, but I stuck and gradually came the ability to endure. At night, when the chores were done, I would sit at the barn and listen to the ducks and the snipe in the Pipestone marsh; then I would close my eyes and hear the bus horses in Piccadilly, and I was very homesick. Very much afraid, too, that I had made a fool of myself with my rustic idealism. However, I could not go back to the gang I had left and admit I was "licked," so I had to see it through.

Presently came harvest and stooking, a heavy, tangled crop, and it was tough to stook. My hands slivered to pieces, and I suffered plenty. After that, threshing, and the cold mornings and the chilly evenings, and I had no underwear. In one way or another, I imagine I got all the misery that could be got out of the situation. I was paid fifteen dollars a month for the first month, twenty-five dollars thereafter. Harvest wages were $2.50 to $3.50 a day, but I was working by the month.

I don't think anyone ever worked any harder for his money than I did for that twenty-five dollars a month. One day I was working at the job of unloading wheat, which came away from the separator in two-bushel sacks, into a large granary, thirty by forty. I packed in one day 1,100 bushels of wheat and walked in wheat to my knees to do it. In my spare time, I figured out I was handling thirty-three tons of wheat a day for one dollar, or three cents a ton—and I did not feel I was overpaid or getting along very fast in Canada.

Every chance we got we were asking about the homestead lands out west. Indian Head was a long way west in those days, but the native boys were taking up homesteads around Eyebrow and Tugaske, which was the wilderness, and few people had been farther west than that.

Halfway through, I made a deal with Fred Reinhart that the day he was through threshing I was through with him, and I stuck to that.

Fred wanted to go duck hunting the next day with some Virden friends, and he owed a team for hauling grain to his neighbour. He wanted me to take his place. He got up to five dollars a day.

"Fred, I'm through," I said. "Make it fifty dollars if you like, but I'm not staying. I'm going into Virden tomorrow."

Reinhart then pleaded that he would need some time to arrange payment of the outstanding wages.

"Fine," I said, "I've learned a thing or two. I'll be staying at the Virden Hotel at your expense. Take as long as you like."

He was there the next day and paid me off—fifty-five dollars—for the fall. That was, I think, the toughest fall that I had.

ENDNOTES

1. The single-stack, 10,757 ton *Virginian* had been in service only since April 6, 1905.
2. Clifford Sifton, Minister of the Interior in Prime Minister Wilfrid Laurier's cabinet, had recruited a large number of immigrants from Galacia and Ruthenia in eastern Europe.
3. The legislation creating the two new provinces had passed the House of Commons on July 5, 1905, and was before the Senate, where it would be approved on July 18. It was to become effective on September 1.

CHAPTER 3
—————

GRAIN BUYER

Their first season of Canadian farming had been harsh, but in October it was over for the year. Charles and Ben wandered down to McClusky, a just-established village noted for being the geographic centre of North Dakota, where a school friend, George McFadden, had a general store. There they found jobs with the Great Western Elevator Company building a new elevator, working for twenty-five cents an hour for a nine-hour day. They declined the offer of an additional twenty cents an hour to join the crew working on the cupola eighty feet in the air. Working on the ground at $2.25 a day was a very satisfactory wage, but their employment ended at Christmas.

I got a little job with the elevator company to watch the lumber which was spread about the site and prevent it going out to the homesteads. The company built me a little shack and installed a pot-bellied stove in it, and there I was to spend the winter, not intolerable by any means.

I formed a friendship with a young German boy who was operating an adjoining elevator for the Great Western Elevator Company, and we moved in with him, and the three of us batched the winter in the elevator office. The young German, Carl Schauer, was quite a cook, and for breakfast he mixed up a pancake batter which consisted of about 50 percent fresh eggs and 50 percent flour and made the most wonderful

Grain elevator under construction in the Prince Albert area, 1906.
Saskatchewan Archives Board photo R-B1765-1-6.

pancakes. In the evening, we had large steaks. They were within our means at the time.

So the winter passed very pleasantly, indeed, and I acquired at least a little knowledge of the art of buying grain, which came in useful later on. Lloyd and I unloaded coal, worked in the livery barn, and did odd jobs of one kind and another at which we made more than enough to eat, anyway. We unloaded coal at ten cents a ton and made $2.00 a day.

With the return of spring, the company started construction work again, and I stayed with them until September, working all over North Dakota.

The standard grain elevator being built across the west, in the United States and Canada, was thirty-one by thirty-three feet wide and eighty feet high. They consumed a lot of lumber. Designed to hold some 35,000 bushels of grain, perhaps 800 tons in weight, they were constructed to be of great strength. Set on a concrete pad, the walls were built of 2 × 6 planks spiked on top of each other horizontally. This wall was lined on the interior with 2 × 4 boards. A competent crew could erect one of these buildings in two to three weeks.

A large lean-to shed was attached to the wall opposite the railway to receive the grain. Wagons were driven up a ramp into this shed, above "the pit," a below-floor hopper tank into which grain was emptied. Then it was elevated by a bucket chain and fed into the chosen bin according to type and grade. Later, when a rail car was "spotted" beside the elevator, the grain was directed into the car by gravity through an external spout.

The elevating machinery was powered by an engine separated some twenty feet or more from the building for fire protection. It was placed at ground level below the small office of the grain buyer, which was connected to the elevator by a walkway containing the drive belt. The engine was frequently a single-piston, gas-fired machine, a "one-lunger" whose slow and regular "chuff—chuff—chuff" beat could be heard throughout the village.

In July 1906, Ben Lloyd returned to Saskatchewan to look for a homestead, but Charles continued building Great Western elevators across North Dakota that summer and kept a record of his travels — Mercer, Turtle Lake, Buttzville, Leonard, Wheatland, Nome, Embden, and Hatton. All his life he carried a vivid recall of the work at Buttzville.

Half a dozen of us were sent to renew the foundations under the elevator which at that point had been in operation probably for thirty years. When we got the lumber at the bottom of the elevator cleaned off and the job exposed to view, we found that the old foundations were festooned with hanging bunches of grain dust six inches in diameter, a most forbidding-looking job. However, we undertook it. In 1906, my status in society was merely that of a common labourer, and I was not expected to balk at any job with which I was faced.

We had to mix new concrete in big mortar boxes, there being no such thing as a concrete mixer in the picture at that time. Then we had to load the new concrete into big pails and carry two of them under the elevator and pour them into the new forms. It was very heavy labour and had to be performed with bent backs.

We discovered that grain dust had qualities all its own. It was capable of penetrating your clothes and reaching your skin all over, and coming into contact with perspiring skin it got well into your pores and caused your skin to swell in patches as big as the palm of your hand, thus causing extreme irritation and annoyance.

Buttzville was a little hamlet consisting mostly of a single country store, and remedies for what was wrong with us were not obtainable. We used to purchase lard by the ten-pound pail and sulphur in generous quantities, mix the two very thoroughly, and apply them all over our bodies, which gave us some degree of relief. Nonetheless, it was a highly repulsive job, and I shall remember it as long as I live. However, it only lasted for a week, at the end of which time we were restored to the sunlight outside and to new construction.

The miserable project at Buttzville was followed by a pleasant experience, equally well remembered. Buttzville, today a ghost town, was located about five miles northwest of Lisbon, in the southeast corner of North Dakota.

The last Sunday we were there we borrowed the hand car from the railway company and went down to the very pleasant little town of Lisbon and headed for the Cheyenne River, a very attractive little river. I immediately undressed and entered the water. I was in such miserable shape from blisters on my body that I did not care if I ever came out. At the time, I was in excellent physical shape and had been well accustomed to swimming a mile in England. I made up my mind to swim a mile in that lovely little river, went up half a mile from the beach first, then turned around and made the mile downstream. None of the locals had ever seen a man swimming a mile, and before I was halfway through I had most of the town of Lisbon for an audience. They got really interested in me, figuring that I was certain to drown, and were yelling at me not to be a damn fool but to come ashore. I was enjoying the physical comforts of the water, which was nice and warm, and did not listen to them, but they kept telling each other, "That damn fool figures he can swim a mile, and we're sure he's going to drown." Of course, nothing happened, and I came ashore in due course to find that I was an object of universal attention, and even admiration, for the feat which I had just performed.

I was certainly in a lot better shape physically, my body was lean, and I had no more trouble with my skin. But I would not like to go through that experience [at Buttzville] again for any consideration, although the length of my life convinces me that I took no permanent damage from the whole episode in any event.

I stuck with the elevator job until fall, then returned to Canada. At that time, several American grain elevator companies were constructing elevators in Canada, one of them being the British American Elevator Company. Before I quit in North Dakota, I got in touch with them, and they told me to come along to Winnipeg, there would be no difficulty about a job. I turned up in Winnipeg, and on Saturday morning I went around to their office for my first interview. It was my idea that I would stay in Winnipeg over the weekend and go to work on Monday. Their ideas were different. I landed in their office at ten o'clock Saturday morning, and at noon I was on the Canadian Northern train headed for Kamsack, my first appointment.

I reached Kamsack at midnight and found I was linked up with a little crew of three husky Swedes, very nice fellows at that. Our job was to unload five carloads of lumber and a carload of cement, build a little shack to shelter the cement, and then go on to another town and repeat that performance there.

I worked at this until fall. We unloaded lumber at a lot of towns between Kamsack and Lloydminster, the names of some of them I have since forgotten. It was a strenuous life. The passenger train travelled those parts in the night, and you had to be ready for the experience of being set off at a bare siding at three or four o'clock in the morning and find your own sleeping quarters in a hamlet where there were no sleeping quarters.

I still remember stepping off the train at Paswegan and wondering what to do about it. There was a boarding car on the siding that had a light in it, and we headed for there, finding it occupied by four Englishmen who were the section crew. We sought shelter with them, but they would not permit us to enter the car.

I went back and consulted with my Swedes and told them there were four of us and four Englishmen there, and that was enough force to enable us to enter that car. But the Swedes refused to follow me, and we had to sleep out in the open at the end of the month of October. A pretty vigorous life, but we survived it.

In the morning, we found a boarding place in town, and before the day was out we had unloaded one of our own boxcars and had shelter for the next night. After Paswegan came Muenster, Aberdeen, Borden, Paynton, Maidstone, and Lloydminster.

I was working on the construction crew at Maidstone when I received orders from the company to return to Dana and open the elevator there. I did this, and the next morning at Dana the doors of a brand new elevator were opened, and I was standing on the scale as a grain buyer, stretching a little knowledge, and it was very little, which I had acquired from Carl Schauer in North Dakota to qualify me as an experienced grain buyer.

Well, the company stood for it, and I got by. The little village of Dana then became my home for two years.

It was the beginning of the golden age of the grain industry in western Canada. Elevators were springing up along every new rail line. In 1900–01, there were an estimated 420 elevators operating in Manitoba and the North-West Territories that became Saskatchewan and Alberta. By 1910, there were 707 in Manitoba, 285 in Alberta, and 1,004 in Saskatchewan, a total of 1,996. Staffing all these new elevators created great employment opportunities.[1]

Operating a grain elevator was work very different from building one. When a farmer drove his team and loaded wagon up the ramp into the shed and onto the scale, the business of grading the grain began. The buyer thrust his hand deep into the load and brought out a fistful sample of wheat. He peered at it, smelled it, poured it back and forth from one hand to the other to see how easily it flowed, a test of dryness, squeezed a kernel or two between his fingernails, and then bit a few kernels to see how hard they were, and applied other personal and equally scientific methods as he attempted to assess if it was Number One Northern, the best, or some lesser grade. While he did so, a usually good-natured banter was carried on between seller and buyer intended to provide more information about the product.

Finally a judgment was made, discussed, debated, and perhaps agreed on. Then the loaded wagon was weighed before the grain was deposited and more imprecision entered the transaction. Grain was priced and purchased by the bushel, a unit of volume measurement. Wheat of good condition and quality weighed sixty pounds per bushel. High-quality wheat could weigh a few pounds more, low-quality wheat a good deal less. The grain buyer was equipped with a scale balance with a small container attached. Weighing the carefully filled container gave the weight of one bushel of the wheat. The balance, and the floor scale, were the only reasonably accurate instruments available to determine the quality and amount, and resulting value, of the grain.

A hatch in the rear of the grain wagon was opened, allowing the grain to spill through the floor grate and into the hopper below. The farmer would clamber into the wagon box with a scoop shovel to encourage the grain to flow and to clean the corners. When the wagon was empty, it was again weighed, the number of bushels and the price were calculated, and the grain buyer completed a cash purchase ticket, negotiable at any bank and most places of business.

What a peculiar thing it was to install me in the elevator, saying, "These are your cash tickets," then hand me $20,000 and tell me, "You are also paymaster." I got a great big canvas pocket sewn into my vest, and for three years with that company I was carrying $20,000 of theirs, and they didn't know one damn thing about me. How would they? I hadn't a root in Canada. Of course, I had a bond, but the only evidence they or the bonding company could have about me was that I didn't appear to have ever been in jail. What else could they have? And to hand $20,000 to a green kid like that? I was twenty-two maybe.

The first were two very trying years indeed. The first crop was fair enough and nearly all dry wheat. The second crop in 1907 was a real problem as it froze while still in the blossom stage. The frost early in

August was accompanied by snow, which made a real mess out of it. There was no known method of drying wheat at the time, and you just had to deal with it as it was.

Conditions were very tough that year, and I can remember being credibly informed that many farmers were living on boiled wheat, and that was wheat that had been frozen just after it passed the blossom stage.

I can remember wagon loads of wheat being hauled into my elevator, and when I got up and walked on them I left footprints in them. This led to an eternal argument with the farmer in respect of grades.

I can remember one outstanding incident where a farmer from Bonmadon, fifty miles north, arrived in my elevator with about twenty sacks of wheat on his sleigh. A large percentage of the wheat in those days came to the elevator in sacks. It was cold, and you could easily freeze your fingers when opening those sacks.

I opened one of those sacks and saw immediately that I had a problem. The wheat had barely passed the blossom stage when it was frozen, and it never reached a better status than that of bran. Flat, shrunken kernels, and they were all coated with smut.

I did a little calculating and then told the farmer, "Look here, I have to grade that wheat Number Two Feed, and there is the price of Number Two Feed. Then I have to discount it for smut, which you can see as well as I can, and further discount it for being damp." I needed no instrument to convince me it was damp. Merely thrusting my hand into the wheat was quite sufficient.

At that time, Number One wheat was selling in the elevators at about fifty-eight cents a bushel. When all these discounts had been applied, I was able to offer the farmer nine cents a bushel for his load of wheat.

He refused to accept this and told me he would dump it in the burrow pit across the road before he would take it. But first he would haul it around town and endeavour to dispose of it for pig feed. I am not quite sure what disposition he made of it. He drove it out of my elevator, and I forgot about the incident.

If a farmer refused to accept the grade offered by the grain buyer, he was free to leave the elevator and try the competition, and some did. A few years after Charles left Dana, he met one extreme example.

There was the case of George Hollenback hauling a load of flax from Section 3–4–5 W3rd to Moose Jaw, 140 miles. The price he was offered made him mad, so he hauled it home again and then took it to Hinsdale, Montana.

Few farmers could afford the luxury of such independence.

So little difference in grading from one elevator company to the next created the strong belief that grain firms were in cahoots against farmers. There were other complaints. Not infrequently a grain buyer would agree that the wheat on delivery should grade Number Two but claim that his bins for that grade were full and that the best he could offer was to take the grain in at Number Three. These concerns grew into manifestations of agrarian unrest and the development of farmer-owned grain-buying facilities.

Few farmers could afford their own threshing machines, the huge separators that predated the combine, and relied on travelling threshing crews, the custom combiners of a later era. If the crop, sheaves standing in stooks in the field, had met with rain, or even snow, it could be unsuitable for threshing when the separator trundled onto the farm. The threshing crew, naturally reluctant to remain idle, could insist on going to work or threaten to move on. It was a difficult situation for the farmer, and damp, or tough, wheat was often the result.

I was clear of that farmer's crop, but in my inexperience I accumulated a lot of tough and damp wheat. As far as I can recall, we had no equipment for determining the amount of moisture in wheat in those years. You simply trusted to your hands, thrusting them into the pile of wheat and judging yourself on what your hands told you. You were under pressure all the time to take the wheat, and it turned out I took a lot of it, to my own grief later on.

Our railway [Canadian Northern Railway] was then operated by McKenzie and Mann, who knew as much about railroading as I knew about grain buying. They had no equipment, no snowplows, no decent locomotives, as they had acquired what locomotives they had by purchasing cast-off locomotives from American companies. It proved to be a very tough winter, and when the snow really came McKenzie and Mann were not equal to the job of keeping the railroad open. They simply threw up their hands and let it go until spring.

In the meantime, my elevator was rapidly filling up with questionable wheat. But, since I had no opportunity to ship carloads of it, nobody found that out until spring. Then the whole tragic story came bare, and my company was very unhappy. I have no doubt they would have dismissed me had they known the facts at an earlier date. I lost grades on everything from the cupola to the pits.

I had made up my mind that I would not again buy grain at Dana for any consideration.

My superintendent had the same opinion. So, in the spring, we parted company, and I became a simple homesteader.

ENDNOTES

1. Brock V. Silversides, *Prairie Sentinels: The Story of the Canadian Grain Elevator* (Calgary: Fifth House, 1997), 10.

CHAPTER 4

HOMESTEADER

During my two years of residence at Dana, I had formed a friendship with one William Devlin, an excellent carpenter working in the village. Willy had undertaken to find me a homestead and did so. Some weeks later he advised me to go to the Land Office in Humboldt and cancel the entry on the SW 22-38-26 W2nd. I was able to do this and was successful in securing a filing on the quarter.

My elevator company, when the construction of the elevator was complete, had a pile of scrap lumber remaining over. With an eye to a homestead shack, I purchased this lumber from them for the sum of ten dollars and had it hauled out to my homestead. Then all winter, every Sunday, I walked out to the homestead, often through three feet of snow, and built my homestead shack with my own bare hands. Of course, I had a little carpenter experience on the construction crews I had worked for, and I managed to produce a small, square building, twelve by fourteen in its dimensions, which stood up. The studs in it were 2 × 4s. They were coated with shiplap and then with tar paper and lap siding. When I had installed a stove in the interior, I felt that I had a home. It, at least, kept most of the winds outdoors, and I can remember standing in my doorway looking over my quarter section and telling myself that I had more land than my father ever had in Ireland.

I felt that I was well on my way. I acquired a nice pony, Prince, and a second-hand buggy and started to do my homestead duties by sleeping on the homestead while operating the elevator in the daytime.

Charles was required to reside on the land for at least six months in each of three years. This turned out to be an extremely harsh obligation to impose on undercapitalized homesteaders. Living on the land before it had been made productive meant flirting with starvation in many cases. Off-farm employment was not readily available. It was a much-repeated joke that the ten-dollar filing fee was a bet against the government's 160 acres that the homesteader could survive on the land three years without starving. The homestead three miles west of Dana gave Charles a huge advantage because he could commute to his job at the Great Western elevator.

Apparently Charles gave little or no thought to why the original homesteader might have walked away from his claim. In later years, he became very attuned to the quality and productive value of farmland, but he failed to perform any diligent examination of his own homestead. He chose it because it was available and reasonably close to his work.

One afternoon I drove out to the homestead, and when I came within sight of it I thought a large flock of sheep were sleeping on it. When I reached my shack, which was well protected by a fire guard, I saw another story. My sheep were boulders. A prairie fire had gone over the quarter during the day time and caused every one of them to stand out against the evening sun. I can still recall vividly the shock the discovery was to me. I put the horse in the barn and took a walk over the land and decided that no sane man would endeavour to clear those stones and break that land.

But, in the spring of 1908, the now-unemployed and determined young Irishman did set about clearing away those boulders and breaking the sod on that land. A hired man was found, and later a team of oxen purchased, and they set to work. Historian Bruce Peel has described the breaking process:

First, the homesteader struck out the land. He erected stakes in a straight line down the field as markers. In working in a half-mile field the ploughman steered his outfit towards the markers.

No sooner was the frost out of the ground in the spring of the year than the homesteader began to break land. He continued to break until the end of June or early July. After that date the sod was too dry. The plough used for breaking had a special share called a breaker bottom.

Breaking was from three to six inches deep. In the fall of the year the breaking was disced, and the following spring the first crop was planted.

Another method of preparing the land for its initial crop was known as backsetting. The land was broken early in the season before the June rains were over. When the sod had become thoroughly rotted by the rains and hot sun, it was ploughed two or three inches deeper. The backsetting was done in August or September. The land was then harrowed. Backsetting was considered by the agricultural experts of the day to be a better method of preparing the land than ordinary breaking. Homesteaders, anxious to bring the land into grain production as cheaply as possible, seem to have been prone to use the first method.

The thrill of owning 160 acres of land made the homesteader's blood tingle, filled him with ambition, and gave him airy visions of future prosperity. He would pause to rest his oxen, lean on the handles of the walking-plough, and look back. Behind him stretched the furrow, and ribbons of black sod, each ribbon resting obliquely on the ribbon turned out of the preceding furrow. The moist, up-turned sod was black, but the hot sun would quickly change it to a grayish brown. The homesteader visualized acres of golden grain, and mentally calculated the financial returns the grain would bring.[1]

I put those boulders in piles with these two hands and the assistance of a stout hired man. We started work early in the morning, and at noon we adjourned to the shack, where I cooked three fried eggs for each of us accompanied by strips of bacon and occasionally by potatoes, if we could find time to cook up a pot of potatoes.

One day my excellent French hired man came to me and said, "Mr. Wilson, I just cannot work any more unless I have some meat."

I said, "I understand what you're talking about. I feel the same way myself. I'll drive over to Frank Wirth's ranch and see what I can do about that."

I acquired a hind quarter of beef, and we kept it in close proximity to the hand axe which we used for a carving knife. Whenever it was getting near mealtime, I hacked off some chunks of beef, regardless of the shape, put them in a skillet, poured water on them, added a few potatoes if we had them, maybe a few turnips, and in two hours we would have a most delectable stew, which certainly made an appeal to men that were working at the heavy labour we were.

We finally completed the job of rock clearing in 1910. We found quite often that, when we had dug these boulders, we had created such a grave in the prairie that on coming along later with four oxen on the breaking plow the oxen refused to cross the depression, and we had to shovel sods into them in order to let the furrow go through. However, we broke that land from line to line.

In late summer 1907, a pleasant surprise came in the form of a letter from Thomas Wilson, Charles' younger brother by one year. Tom had followed Charles to Canada and was in Winnipeg. And broke. He came to Dana and then went on to Prince Albert and to the northern lumber woods, where he worked for the winter. He returned in the spring of 1908, and the two brothers began homesteading and farming, *not knowing anything about either one,* as Charles freely admitted.

They were able to secure for Tom a homestead entry on the adjoining quarter section, SE 22-38-26 W2nd, when the original homesteader there also decided to throw in his hand. *With the land came quite a large log shack, he had a family, three or four rooms. And then he had ploughed up the sod on the prairie, and he built sod over it until it looked like a beehive, but it was warm, fine in winter and fine in summer. You could be in there and have a thunderstorm, and you wouldn't hear it.*

We moved from mine down there and did the rest of our homestead duty there in comparative comfort. We didn't freeze, anyway.

The interior decor of their new home was rustic but functional. The partitions were built of bark-sided slabs that looked very much as if they had been purloined from the snow fence along the adjacent railway.

Charles and Tom now had a full half section of land, 320 acres, four times larger than Driem, their father's farm back in Wicklow. It was the beginning of a successful partnership between the two brothers.

In 1908, I was on the homestead and no more interested in political matters than any other hard-pressed homesteader. I had, however, in the elevator at Dana accumulated a wide personal experience with Galacian people, so many of whom had settled around Cudworth and Wakaw. I was indebted to my friend and neighbour, Frank Ralph Wright, that summer for about $500 for oxen, and one day Frank approached me and asked me to canvass the Cudworth and Wakaw communities on his behalf. He was running as an Independent Liberal, being dissatisfied with the Liberal convention, which had failed to nominate him and had instead handed the nomination to one A.F. Totzke, a druggist in the town of Vonda. I thought the debt between us was a sufficient reason for complying with his request, and I agreed to do so.

Walter Scott, a Liberal, had been designated as Saskatchewan's first premier in September 1905 and had been confirmed in that office by a general election in December. Now, not three years later, Scott called a provincial election for August 14th.

Frank Wright bought me a suit of clothes for the sum of eight dollars, without which I had no clothes to make me presentable to the public. The next morning he came to the hotel and told me the team was ready. I went downtown with him and found the team was a pair of wild broncs who were hitched up for the first time in their lives. They were straining at the harness while four men held them. Frank told me to get into the buggy, and on the other side he put in one Alec Kozack, whose capacity was that of interpreter.

As soon as we were firmly seated in the buggy, the four men withdrew their hands from the team, which started off on a wild gallop. We had a winding bush trail to face on the road to Cudworth, and for the first five miles we yelled our heads off in the hope of warning any southcoming traffic, but fortunately we met none.

I was amazed at the quickness with which that team picked up the facts of life. By noon, they had cooled off a great deal, but just the same we did not dare unhitch them. We simply took the bridles off and fed them, still hitched to the buggy. Of course, they continued to get tired and more tired every hour, and presently we had quite an easy team to handle. They had noticed that we stopped at every thatched house, the Galacian people of that period all having the same style of roof. One afternoon the team left the road and turned into a substantial, well-thatched building which turned out to be no more than a granary erected by a farmer, one Otto Ziegenhagen. By the end of the week, the team was thoroughly docile, and, indeed, Friday afternoon we had to get hold of a willow limb to serve as a whip.

We returned safely to Dana and went out election day to a poll north of the present village of Prud'homme. I knew nothing about the Election Act, although I had one in my pocket, but that wild team effectively interfered with any detailed study of it. At this poll, we were faced with an experienced DRO [deputy returning officer] and candidate's agent.

When they started to come into the poll for the purpose of voting, I demanded that they be sworn. Immediately a difficulty arose. The DRO had neglected to furnish himself with a Bible, and so we had no means of administering the oath. He first produced a book, a large notebook issued by the T. Hamm Brewing Company of St. Paul, Minnesota. This I knew enough to reject, and now we were stuck. Finally we compromised on a Galacian prayer book and used it for the balance of the day. By this time, practically every elector at the poll had come in and gone out without voting. I did not know enough to know that that fact precluded their re-entry to the poll, and later in the day I allowed my opponent to bring them all back in and vote them, with the result that we lost the poll very badly. However, as a result of our efforts in the Cudworth and Wakaw country, we came very near to winning the election in any event, and it does not, at this stage, matter one little bit that we failed to do so.

Premier Scott and his Liberals won twenty-seven of the forty-one constituencies, while Frederick Haultain and the Provincial Rights (later Conservative) Party won fourteen. Frank Wright lost the election but ran a very respectable second to the winning Liberal, 321 votes to 442. The Provincial Rights candidate ran third, and last, with 174 votes.

Charles lost the first election he was involved in, the first of many to follow and the only one in which he opposed the Liberal candidate. The Irish in him had tasted the brew of politics, and he remained addicted until the end of his days.

The British American Elevator Company did not give up on Charles as a grain buyer, even after the costly experience at Dana, and it posted him to Snake Creek, Manitoba, and then to Angusville for the winter seasons of 1908–09 and 1909–10. Tom stayed on their homesteads to keep them going while Charles, out of his monthly salary of sixty-five dollars, sent him twenty-five dollars as a contribution to shared expenses. The summers were devoted to clearing the never-ending rocks and breaking sod in preparation for their first crop. The rock piles that Charles created were so impressive that thirty years later, long after he had sold and left the land, he returned to display them with pride to Chief Justice J.T. Brown of the Saskatchewan Court of King's Bench.

On my return to the homestead in 1910, we completed the breaking and seeded a crop of wheat of over 100 acres, from which we expected great things. 1910 proved to be an extremely dry year, and we had a real crop failure. I still remember the afternoon that the threshing machine pulled off our place, having threshed a total of 600 bushels of wheat for us. Out of this had to come seed, and we had debts all over the place. We had a particularly large one in the general store, then operated by Louis Normand, of Dana, one of the greatest gentlemen it was ever my good fortune to meet.

After the threshing machine pulled off, I took the oxen to the pasture and turned them loose and leaned over the fence contemplating our prospects. I then went into Dana and paid a call on Mr. Normand and told him frankly our circumstances and that it was out of the question *for us to attempt to pay him the $700 we owed him for groceries obtained during the past two years. I told him, moreover, that I could not see at the moment how it would ever be possible for us to pay him and that he would have to cease supplying groceries to us.*

He sat me down and talked to me as a father to a son and told me that I would continue to come in to his store and that he would continue to furnish groceries and that finally I would pay him every cent I owed him.

This I did. In the fall of 1911, we had a much larger crop on a couple of hundred acres of wheat, and 1911 proved to be an excellent year. We harvested nearly 5,000 bushels of wheat, a very large crop in those days, and we were able to go in and pay Mr. Normand in full.[2]

That fact by no means eliminates the gratitude that I have felt towards Mr. Normand all my life. I remember him telling me that he homesteaded at St. Rose du Lac, south of Dauphin, Manitoba, and that he had gone through exactly the same experience that we were passing through. I remember him telling me that he broke furrows on the raw prairie, depositing potato seed underneath them, and the following fall harvested a luxuriant crop of potatoes on which he existed for a year. I believe that Mr. Normand is buried at Prud'homme, and I still hope that I will have an opportunity of visiting the cemetery at Prud'homme and kneeling at his grave. He was a most valuable friend to me at an extremely critical time of my life, and I hold his memory in deep respect.[3]

That 1911 crop was the last one for Charles on the homestead. Tom had taken to trading in oxen and horses and had developed a nice little business, but the transactions were largely based on credit, an aspect that made his more careful brother nervous. Charles looked for opportunities beyond the homestead. He had secured title to that homestead in December 1910, having fulfilled all his homesteader obligations. Tom completed all his requirements the following August and took title to his quarter section.

The tide of immigration pouring into Saskatchewan lifted its population to just under 500,000 by 1911, making it the largest of all other provinces except Ontario and Quebec, a distinction that it would retain until after the Second World War. Saskatchewan also held the third largest number of members of Parliament, giving it a strong voice in national affairs.

Charles was so concentrated on becoming financially established that even his experience in the 1908 election had not yet fully drawn him to an interest in public affairs. All around him, unhappiness and unrest were bubbling throughout the growing western farm community. The concerns were national. In December 1910, a delegation of 500 western farmers travelled to Ottawa to join with 300 Ontario farmers in presenting their grievances to the government of Prime Minister Laurier. Farmers were then the most potent political force in Canada, and the delegates were permitted onto the floor of the House of Commons, the only time that privilege was granted.

In 1912, the Canada Grain Act was enacted, addressing some of the farmers' concerns, but discontent remained, and within a few years western farmers began to take politics into their own hands. By then, Charles was very alert to the political necessity of meeting the needs of the farm community. Later still he became deeply enmeshed in the tribulations that had befallen western agriculture.

I started to look around for an opportunity somewhere other than the homestead, and the South Country was very much in the news at that time. I left the homestead and travelled to the town of Morse, where I hired a livery team and drove south to Gravelbourg, a three-day trip, but I saw the land, and I liked it, mostly good level land, comparatively free of stones.

I returned to Dana and asked my brother to buy me out. All I had to sell was the SW 22–38–26W2, for which I asked him the price of $2,300, minus my mortgage to Canada Life outstanding at $1,050, leaving me an equity of $1,250.

The $2,300 total price was fine with Tom, but again, as usual, there was no cash in the deal. It was done by a form of security between seller and buyer known as an Agreement for Sale. The legal rights were similar to a mortgage, but Charles retained title to the land until it was paid for.

I took the agreement into Saskatoon with the idea of selling it, and first thing in the morning I approached the office of McCraney, McKenzie, Hutchison, and Rose and offered it to them. They made me a cash offer of $900. I stalled and spent the rest of the day walking the sidewalks of Saskatoon hoping to improve that offer, without success. At four o'clock, I returned to the office and made a deal with them.

That $900 was every dollar left in Charles' pocket after six years of hard work in Canada, the last four years homesteading and farming. The 1911 crop had bailed him out of all his debts other than the Canada Life mortgage. At that, he had done a lot better than many other homesteaders.

Charles returned to grain buying for one more winter, this time at Frankslake, about twenty miles northeast of Regina. When the spring of 1912 arrived, he was ready to embark on a new venture. As yet, he did not know what that new venture would be, but he was heading into the South Country, those mostly new lands lying south of the Canadian Pacific Railway mainline and west of Moose Jaw.

END NOTES

1. Bruce Baden Peel, "R.M. 45: The Social History of a Rural Municipality" (MA thesis, University of Saskatchewan, 1946), 250–52.
2. The price of wheat at Fort William in 1911 was approximately a dollar a bushel. The elevator price would have been somewhat less but sufficient to produce an excellent cash return to Charles and Thomas.
3. Ludovic Normand, born April 2, 1861, at Pas de Calais, France. After studying the English language in England, he emigrated to Canada in 1894 and homesteaded near St. Rose du Lac, Manitoba, where he also served as postmaster. In 1904, he moved to Dana, where he operated a general store and the post office. As the only store between Humboldt and Warman, it did a large business. In 1915, he moved his business to the nearby larger community of Howell, which, in 1922, became Prud'homme. Ludovic (Louis) died at Benito, Manitoba, on November 15, 1954.

CHAPTER 5

TO THE SOUTH COUNTRY

I bought a team and buggy from my brother, the loveliest team of horses behind which I ever sat. They had been raised and broken by one Bill MacPherson of Cudworth, and Billy was a great horseman. This team would do anything you wanted them to, provided you could make them understand what that was. I piled my grip in the buggy and came in overland in May of 1912.

In later years, Charles recounted the most desperate experience that he had with that team, trying to control the high-spirited geldings during a severe hail storm from underneath the buggy, where he was sheltering from the hail and coping with a severe attack of diarrhea.

Charles had a strong faith in the country that he had inspected so briefly the previous fall, but he was running contrary to a good deal of conventional wisdom, though likely he was mostly unaware of it at the time. He was heading into what for many years had been known as the Canadian extension of the Great American Desert, a vast, arid, treeless region with minimal rainfall. It still suffered from the stigma of the Palliser Triangle, land declared by Captain John Palliser in 1857 to be unsuitable for settlement, an opinion confirmed the next year by Professor H.Y. Hind of Trinity College, Toronto, who stated that the area was "not, in its present condition, fitted for the permanent habitation of civilized man."[1]

These pessimistic views of the agricultural potential of the southern lands were largely debunked twenty years later by an appraisal carried out by Professor John Macoun, a naturalist retained by the government of John A. Macdonald, anxious to complete the Canadian Pacific Railway. Macoun toured the west at a time of greater precipitation and was favourably impressed. He reported that some 80 percent of the country was "suited for the raising of grain and cattle."[2]

But, in the main, the southern lands were considered suitable only for cattle. The Department of the Interior in Ottawa agreed with Macoun that the Great American Desert did not exist in Canada and that "in the very worst parts of the country many tracts of good soil were found, and almost invariably the grass was rich and nutritive, offering excellent facilities for stock raising."

Minister of the Interior Clifford Sifton held to the view that the lands lying just above the forty-ninth parallel in much of what became the provinces of Saskatchewan and Alberta were suitable for ranching only. Those lands were not made available for homesteading but reserved for grazing as far north of the border as the mainline of the CPR. A map issued by the Department of the Interior showing the disposition of lands in Saskatchewan up to January 1, 1912, clearly identifies those lands east of the Missouri Coteau as far north as Elbow and only then trending to the west.

But in 1905 Sifton resigned from the Laurier government and was replaced as minister of the interior by Frank Oliver, an Edmonton MP who brought to office a very different view. Oliver thought ranchers to be a landed and privileged class impeding the settlement that the west desperately needed.

In 1908, Oliver secured an amendment to the Dominion Lands Act that opened to homesteading the great tract of land between Moose Jaw and Calgary and south of North Battleford. In consideration of the acknowledged lack of rainfall in the region, an inducement was added. Much of the dryland area now became known as the "pre-emption area," and homesteaders were granted the right to a second quarter section of adjacent land. The government proposed to provide the settler with an additional 160 acres, conceding that a half section would be required to provide a livelihood in the drier region of lighter soils.

Swarming crowd at the Dominion Lands Office in Moose Jaw, 1908.
Saskatchewan Archives Board photo R-A501.

The pre-emption area was a huge tract of country, some 314 township, lying north of the American border and extending from the "Soo Line" of the CPR, which angled from Estevan through Weyburn to Moose Jaw, then west into Alberta almost as far as the foothills of the Rockies. In this region, 50,000 pre-emption quarter sections were sold at three dollars per acre.

The new country became available for homestead filings on September 1, 1908, and an immediate land rush swarmed the dominion lands office at Moose Jaw. When the doors opened that morning, more than 1,000 men were waiting, many of them having been standing in line all night. For the remainder of 1908, a waiting crowd of 200 or more was common, and a total of 8,710 homestead applications were processed. In 1909, 9,573 homestead filings were handled by the Moose Jaw office. In April 1910, part of the district was handed to a new office in Swift Current, and in that one year the two offices together processed 12,133 filings.

The wisdom of the decision to allow the conversion of all that grazing land to grain farming was debated for decades and came into serious question in only ten years. Much of the land was suitable for breaking, but much of it clearly was not. The failure to identify the unsuitable areas was a tragic omission. Many hundreds of the hopeful homesteaders who swarmed into the South Country were heading toward years of hardship, misery, and failure. Large regions in southwestern Saskatchewan and southeastern Alberta suffered almost continual crop failures only ten years after settlement.

Driving his team overland into that new land in the spring of 1912, Charles was intending not to do any more homesteading or farming but to establish himself in business serving that burgeoning population. Even so, his determination was tested the first night of his trek south when he camped with his old friend William Lilwall, north of Colonsay.

Bill kept me up most of the night trying to persuade me to hitch up in the morning and go right back where I came from. He depicted the South Country as a burnt-out country where nobody could expect to succeed. However, I was obstinate and persisted on taking my own way.

I drove south through Colonsay, Craik, Tuxford, and Moose Jaw. When I reached the north end of Main Street in Moose Jaw, I thought, "Now, I'll have to be careful. This team will probably bolt when they meet their first streetcar." So I tightened the lines and got all set. Presently we did meet a streetcar. The team was frightened, and their front legs buckled under them. They went down on their knees, but they never tightened the reins in my hands. I often think of all the miles they carried me in the South Country, and I think of them still with both affection and respect.

I am now at the top of the north end of Main Street in Moose Jaw, and I have a long-distance view of my beloved South Country. I sit there in the buggy and tell myself that surely in that vast expanse is a spot where I can found a little modest home and make a living and have a nice girl across the table from me to pour the coffee. I think that I have realized all these modest objectives.

When I left Moose Jaw, I had $700 in the Union Bank there, forty or fifty dollars in my pocket, my team and buggy and my grip, and nothing else. Except an agency connection with Anderson and Sheppard, Moose Jaw, for hail and fire insurance and an understanding with Canada Life *that they would look the South Country over and probably give me an agency. I had a contract with the life department.*

In 1885, a telegraph line was constructed from the line that accompanied the mainline of the CPR near Moose Jaw to the North West Mounted Police post at Wood Mountain, established ten years earlier. The new line was carried by poles erected along a southwesterly route that ran east of Old Wives Lake, past the yet-to-come communities of Crestwynd, Expanse, Mazenod, and Limerick, then due south to the NWMP post on the northern slope of the Wood Mountain uplands.

The line of poles was a natural navigation aid to the early settlers, and a trail developed, the Pole Trail. The telegraph poles marched in a straight line over hills and through sloughs, while the trail followed an easier route but always in sight of the telegraph line. The trail remained in use for some thirty years, maturing into the Old Pole Trail. When Charles arrived at the trail in the spring of 1912, he found it to be a busy highway.

It surely was an important and historical feature of the South Country. In 1911–12, an average of 300 wagons used to leave Moose Jaw headed for the South Country every day, most of them four-horse loads. A great deal of human suffering took place on the trail, and in 1912 it was fairly well marked from Moose Jaw to Limerick by the skeletons of horses which had died on it. It was also the scene of much human kindliness and neighbourliness. When a man got into trouble on it, due to the death of a horse, the next travellers would stop and pick up part of his load and lighten it for him so that he was able to proceed to the end of the trip.

The Pole Trail, during the last years of its existence, must have been one of the busiest roads in Canada, and almost the entire settlement of the South Country was accomplished over it. It deserves to have a place in any memoirs of the South Country.

Charles headed down that trail in the spring of 1912. Along the trail developed "stopping places," where travellers, or "stoppers," could feed their animals, buy meals, and secure a bed. South of Old Wives Lake (then called Johnson Lake), the Murphy family kept a stopping place, whimsically called "Lakeview Hotel." Their sign read "Pork and beans for man or beast." Another wayside inn had a large picture of hogs feeding with a caption reading "First come, first served."[3]

I stopped for lunch at Ed's Ranch, east of Dunkirk, and there for the first time I made the acquaintance of John A. Maharg [see pp. 227–228], who was in the district selling stock in the Saskatchewan Co-operative Elevator Company, a concern which became immensely prosperous under the general management of Charles Avery Dunning, who, a few years later, turned up as the popular premier of this province.

I spent the night at Expanse and next morning resumed my journey on the Old Pole Trail. By evening, I had arrived at Valor, and it was time to look for a place to stay overnight. I saw a farmer in the field and drove in to interview him. He proved to be Mr. Able James Hindle, later on the Liberal member for the constituency of Willow Bunch, a position which he held for some years. He invited me to his house, and there I was successful in making a deal with his wife to permit me to come into the family as a boarder and to remain as long as I pleased.

That turned out to be a very happy decision for me. It was a lovely family to be associated with, and I want to record that Mrs. Hindle was more painstaking in her care of me than my own mother had ever been. I spent two very happy years under the Hindle roof before moving to Limerick.

In the spring of 1912, we were getting to the stage in the South Country where businesspeople were sticking up signs all over where they estimated the townsites were going to be and announcing their entry into the picture. There had been a little hamlet northeast of Assiniboia, known as Leeville, and at the start all the signs were directed there. Four banks announced their intentions, and there were four real estate offices such as I hoped to open when I got located. That pretty well eliminated me from Assiniboia, and I started to look farther afield and decided there would be a good town at Limerick, where I was likely to have it all to myself. Accordingly I had lumber hauled from Expanse and erected the first building on this townsite, about a twelve by sixteen building in which three of us slept and lived for maybe six months.

Charles had another reason for choosing Limerick. Beginning as early as 1907, before the township survey made actual homestead filing possible, a substantial and growing number of Romanian settlers had located just a few miles south of Limerick, along the north shore of Twelve Mile Lake. Charles had learned during his grain-buying experience at Dana that he had a talent for getting along with the immigrant people from eastern Europe. For many years, he and the Romanians of Sts. Peter and Paul parish, near the later community of Flintoft, enjoyed a pleasant and close relationship, business, personal, and political.

I drove the prairies all summer in 1912 selling hail insurance on notes and earning from Anderson and Sheppard commission certificates for my equity. These certificates were payable when the notes were paid, and the crop had to be harvested and hauled to Expanse or Viceroy before that could happen. A long wait, and I had to make some provision for winter.

Steel was then at Expanse and at Viceroy. There were few roads except winding prairie trails. All wheat east of LaFleche had to be hauled to Expanse or Viceroy, west of LaFleche to Vanguard. In the fall of 1912, the pressure on Expanse was so great that many times farmers went on to Moose Jaw to unload and saved time by so doing.

Expanse had responded to the economic opportunity of having the railway end-of-line. Its business district consisted of two livery barns, two blacksmith shops, five lumber yards, a telegraph station, two hotels each with a café, three Chinese restaurants, a livery dray service, and a dance hall.[4] Today it is virtually a ghost town.

In August of 1912, I got my bankbook from Moose Jaw and noted with alarm that my capital was getting away from me very rapidly, and with winter in the picture it was up to me to do something about it. One afternoon when I returned to the Hindle farm, there was a gentleman, a Mr. Stewart, from the M. Rumely Company, waiting to see me. He told me that they had a very large number of threshing machine claims scattered all over the South Country, and they wanted a man who had a little knowledge of the country, and a team and buggy to enable him to get around, to undertake the collection of these claims. He finally inquired if I would be interested.

In view of the condition of my bank account, I was greatly interested, and he then asked me at what figure I would undertake the job, paying my own expenses.

Mr. Hindle was building a new barn, and there were scraps of lumber lying around. I picked up one of these and did a little figuring on it, trying to look as wise and experienced as possible. Then I quoted him a figure of $200 a month, whereupon he promptly told me, "Come into Regina tomorrow and get your supplies and go to work."

I now felt as firmly established financially as if I had a bank behind me. Within a month, I discovered that the other collectors, who used the railway and livery teams to operate, were getting $250 a month, and their average expense account was over $200. I never raised an objection. I had signed the contract for $200 a month, and I adhered to it.

The contract expired on the first day of December, and, of course, winter was coming about that time, too, so I headed for Regina. When I walked into the manager's office, one Arthur H. Kennedy, a very great gentleman, asked what I was doing in town.

I said, "Where else would I be? My contract has run out, and moreover, even if you renew it, I could not possibly undertake to drive every day in the wintertime."

"We'll soon fix that," he said. "You go back home and keep on working. We'll put you on half salary so, if the day is too tough, you won't feel obliged to go out in it."

I accepted the offer, came home, and continued to work. That proved a lifeline for me. I had a little money every month and was enabled to carry on with my own objectives. In addition, I was rapidly getting acquainted with the people and the territory of the South Country, assets which proved invaluable to me later on.

In the fall of 1914, I was still working for Rumely under contract for $200 a month, inclusive of expenses. I was also selling their heavy machinery under an agency sales contract, and I was producing a lot of business. The company at that time had its branch office in Estevan. The two managers called me into Estevan and faced me with a new proposition. Each of them wanted to keep me on their staff, but only at the price of quitting the other one. They told me I was upsetting all the other agencies in the country by getting around so much and butting in

Rumely steam engine breaking sod. Photo in possession of the author.

on the prospective sales of the local agents. I knew that to be true, and I tried to talk them out of their idea of having me resign from either office, but I failed to succeed. Finally, I told them I would relinquish my sales contract and hang on to the collection contract. It was still paying me the lousy $200 a month for which I had hired on in the first place, but on the other hand it was cash at the end of every month, and I needed cash to pay my board bill.

I also told the sales manager that, while I would surrender the Rumely contract, I was not ready to get out of the business and that I would go to Regina and replace his contract with another one. This I promptly did, and it had immediate consequences.

Rumely had been trying to sell a steam engine to Charlie Burgeson, of Congress, for many years but had not succeeded in closing the deal. Mr. Burgeson took a sudden notion that he was ready to close, and he was kind enough to let me know that he was about to take this step. I had secured a contract from the Emerson-Brantingham Company, and I accompanied Mr. Burgeson to Regina and succeeded in selling him what I believe was the last steam engine that went out of Regina. Steam had become obsolete in favour of the gasoline tractors.

We sold Mr. Burgeson a huge engine described then as a Reeves cross compound with a steam steering device. I have seen this enormous outfit in the field later on with Mr. Burgeson packed in a cab about sixteen feet off the ground, and I told myself then that Charlie Burgeson would rather be up there operating that immense outfit than be president of the Canadian Pacific Railway. The outfit pulled fourteen plows, and with it Mr. Burgeson broke an immense area of the South Country, from the correction line south of Moose Jaw to Congress. The contribution which he made to the swift development of the South Country should not be ignored. The South Country passed from raw prairie into wheat fields in an incredibly short time due to the organization of men like Charlie Burgeson.

ENDNOTES

1. Cited, Owram, Doug, *Promise of Eden*, 1992, University of Toronto Press, 67.
2. Cited, Waiser, W. A., *The Field Naturalist*, 1989, University of Toronto Press, 42.
3. Peel, "R.M. 45," 151.
4. *Leader-Post*, January 10, 1975.

CHAPTER 6

WILSON BROTHERS *and* VICTORIA TRUST

S teel for the Canadian Pacific Railway branch line from Assiniboia to Shaunavon reached Limerick in the summer of 1912, and the townsite was established. Service on the new line began the following year. By August 1913, the community had progressed to the point that it was organized as a village, and a council was elected, consisting of an overseer and two councillors, of whom Charles was one. The first meeting of the new council was August 14, 1913.

The nascent community displayed a strong spirit of cooperation and understanding that extended even to its several religious faiths. The Holdsworth School was moved from the country into the village and became the Limerick School. On Sundays, it was devoted to church services, with the Anglicans and Catholics alternating Sunday mornings and Sunday evenings reserved for the more numerous Methodists.

The maintenance of law and order and the administration of justice encountered frontier problems. On October 13, 1913, W. Brown was appointed village constable, but just a month later members from the Wood Mountain detachment of the Royal North West Mounted Police identified Brown as the culprit in a small crime spree, arrested him, and charged him with burglary.

Charles had been appointed as Limerick's first justice of the peace and was called into service on an interesting case involving a Finnish homesteader known as John the Finn.

One morning he got up early and decided to go on a bender. He showed up in Joe Thompson's blacksmith shop with a bottle and insisted that Joe have a drink. Joe was busy, with a horse's leg up, shoeing it, and he was rude to John, told him to go away, he didn't want a drink. John was insistent, and Joe picked up something and hit John.

John took mortal offence and left in a hurry and came back with a revolver in one hand and a rifle in the other. Joe saw him coming out his window and left in a hurry with his leather apron sticking out behind him like a shelf so fast was he going. He holed up in Mr. Hirsch's cellar.

When John arrived at the blacksmith shop, there was nobody to shoot, so, frustrated, he started looking for trouble. The affair was reported to me as JP, and I made out a warrant for John's arrest.

I thought I had some influence with old John, and I went out looking for him, thinking I could induce him to give me the guns. Instead, he pointed the rifle at my stomach, the only time in my life I had that experience. I confess that my stomach turned to jelly, and I got out of there in a hurry and engaged in the same occupation as my neighbours, trying to keep at least one, and better two, buildings between me and John, who roamed the town for an hour or two and dominated it.

I went back to the office and wired Wood Mountain for a policeman. My hand was very shaky, and that telegram was hardly legible.

But I got by with it, but everybody was coming in, saying, "That man should be arrested." I'd say, "I agree with you. Here's a warrant. I'll make you a special constable. Go and arrest him." Nobody would go.

Finally a big Irishman of the name of Martin Lynch, who had a restaurant in town, came in and said, "Give me that warrant." I did so, and in about fifteen minutes Martin returned to me with the two guns in one hand and John the Finn in the other. John surrendered, and we had a little jail and placed him there, waiting for the mounted policeman to show up and straighten the picture out. Because everybody rather liked old John, he was only charged with one offence, for which I gave him sixty days confinement in the Wood Mountain barracks, which he served.

While he was there, the inspector for the Wood Mountain detachment went out of his way to tell him that I had no authority to add the injunction, which I had done, that he should not return to Limerick, and at the end of the sixty days John came back and started to frequent *the bar, where he told the boys that he was going to stay in Limerick long enough to plant a knife between my shoulders.*

A bootlegger friend of mine came in one evening and warned me to take John seriously, that he really meant it. I said, "All right. I'll meet him head on." I got hold of the village constable and sent him up to the bar to re-arrest John and bring him down, which he did. I then told John that, while he had committed twenty-five or so offences while he had been on his rampage, out of consideration for him he had only been charged with one. And I added, "If you don't leave town right away, you are going to be charged with the other twenty-four."

John took me seriously too, and next morning he started out to walk to Assiniboia. I had occasion to go to Assiniboia with a passenger, and I overtook him about halfway there, plodding along. I stopped and offered him a ride, not knowing until I stopped who he was, and put him in the back seat of the Ford car, giving him the finest opportunity in the world to carry out his expressed intention. However, nothing happened. I set him off at Assiniboia, and nobody in Limerick has ever see him since, but he sure upset this community for half a day. In spite of their hurry to get their buildings up, nobody did any work that morning.

Charles soon became known as "Irish Charlie" Wilson, and many assumed that he had contributed the name Limerick to the new community. But the name had been chosen by one Edward Lossing for the telegraph key on the Moose Jaw–Wood Mountain telegraph line that was installed in his home in 1907. The name moved to the later post office and then to the village.[1] The streets were also given Irish names: Galway, Connaught, Shannon, Killarney, Kerry. It was pure coincidence that the Irishman from County Wicklow found himself living on Kerry Street, Limerick, for more than fifty years.

Charles established himself in business as a mortgage broker and local insurance agent, offering farm loans and fire and hail insurance. Business was immediately brisk, with a great demand for farm mortgages:

Without the availability of vast amounts of credit the rapid development of the prairies would have been impossible. From the earliest days of settlement the people have operated to a very considerable extent on funds made available from outside sources. During the homesteading periods advances were made by lumber yards, implement companies, stores and banks, and usually when the duties of homesteading had been completed, the different loans were consolidated in a real estate mortgage. As the districts developed, additional lands were acquired by the homesteaders, usually on the deferred terms of agreements for sale.[2]

In 1913, my brother Tom discovered that operating the farm at Dana involved a lot more labour than driving around the road allowances dealing in oxen and horses. I was aware of this and invited him to come down and join me. He came with alacrity, and so we founded the firm of Wilson Brothers, Financial and Insurance Agents, which proved to be quite an enduring institution later on. I had the best business partner in my brother of anyone in the province. He was highly intelligent, ingenious, and personable and could get over any difficulty with which he found himself faced in short order.

The country was crazy for new capital at that time. We had practically every Canadian mortgage company represented in the office, and I think we were carrying on about the most extensive farm mortgage business in the entire province. The business was coming to us from the right and the left, from the front and the rear, and we were frequently hard-pressed to keep up with it, although at one time we had a staff of five in the little office at Limerick. My brother remained in the office and kept things going there. I got up early every Monday morning and disappeared into the remote stretches of the South Country, contacting business everywhere and securing new clients. In spite of this activity, we reached a point where we were incapable of handling all the business that was offered to us, but we did establish a reputation for the fact that, if you could not borrow money in Limerick, you could not borrow anywhere. This was a solid asset, and we tried to procure money in every direction. My brother made two trips to Minneapolis looking for money and did succeed in interesting Drake, Ballard and Company in the field, and from them we secured a lot of money, even if later on it proved to be very high-

priced money. I can remember it was repayable in gold, and some of our farmers, when it came time to pay the mortgage out in full, were stuck with premiums as high as 22 percent in order to return the money to Minneapolis.

In the fall of 1915, a Calgary stenographer came here to visit some friends. I managed to induce her to come to work in the office. She was in a class by herself, able and competent and a perfect stenographer. She came in to help on a temporary basis and remained for the rest of her working life—Miss Elizabeth

Charles and Thomas, 1915,
the "Wilson Brothers."
Photo in possession of the author.

(Lizzie) Allonby. She was very easy to get along with, extremely competent, and in six months' time she had taken control of the office, reorganized it, and was acting as the general manager. I never worried about any matter being left to her decision. She had experience, and her judgment was to be trusted at every turn of the road.

The year 1915 brought another piece of good fortune to Charles, a connection to an Ontario loan company that grew into a fifty-year relationship of unusual strength and warmth. One of the Romanian settlers south of Limerick came to Charles with a problem: a judgment had been registered against him by another Limerick agent for a matter supposedly on behalf of his farm loan lender, the Victoria Trust and Savings Company, of Lindsay, Ontario. When Charles took the matter up with Victoria Trust, C.E. Weeks, the managing director, was horrified at the malfeasance supposedly carried out in the name of his company and requested that Charles correct the problem, which was promptly done.

Victoria Trust, a young but growing corporation, had been founded at Lindsay twenty years earlier by R.J. McLaughlin, KC, a lawyer and a paragon of rectitude who ensured that the company's officers and employees shared his high standards. After Charles cleared away the initial difficulty, Victoria Trust, wanting to increase its farm mortgage volume in southern Saskatchewan, appointed Charles its loan agent. That simple beginning blossomed so that Wilson Brothers, of tiny Limerick, became the provincial manager of all Victoria Trust's holdings in Saskatchewan.

Canada Life Assurance Company was the most active mortgage lender that Wilson Brothers represented and placed a great number of loans in the South Country. More than $2 million of Canada Life loans were placed through the Limerick office, representing almost 1,000 loans, since those early mortgages were usually secured on only one quarter section of land and averaged $2,000 to $3,000. Charles enjoyed a close relationship with the management of Canada Life in Regina.

We had literally a gold mine here, and we worked as assiduously as possible. This went on until the spring of 1916. At that time, my brother visited Winnipeg for a weekend and returned from the trip in uniform. The First World War had overtaken us. I did not buy him out; I simply undertook to carry on the business in which we were jointly interested as effectively as I was capable of.

It had been a year of bumper crops all over the prairie provinces, with wheat yields exceeding twenty-five bushels per acre. The volume of production brought the price down from $1.62 in the spring to $1.00 by the fall but still resulted in a huge economic boost. Production fell off in 1916 and 1917, but prices soared, reaching $2.21 and then $2.63 by 1919. There was a great demand for farm mortgages.

Most farmers were in a hurry to have their farms fully equipped. Their confidence in the country untried by drouth, many farmers thought it safe to mortgage their farms. Wishful thinking made it "next year" country. Farmers convinced themselves that they must be ready to exploit a good crop year by farming on a more extensive scale. Hence they must buy expensive machinery on credit. They found the loan companies and banks anxious to grant loans, and machine companies willing to sell on the time-payment plan. ... The provincial commission appointed in 1913 to investigate farm indebtedness found that in some districts in the province as high as eighty per cent of the farms were mortgaged.[3]

When Tom enlisted in the Canadian Army in 1916 and shipped over to Europe to participate in the First World War, Charles was left to carry on their thriving business alone. As busy as he was, he found time to marry the woman he had met two years earlier in somewhat unusual circumstances.

In the fall of 1914, I was still working for the Rumely Company, driving a team of ponies all over the South Country looking after their collections. I stayed one night at the J.R. Copeland farm on Section 2-8-5 W3rd. Mr. Copeland was not home, but there was a young Methodist minister in charge who made me welcome and fed me. I was travelling south, and in the morning I noticed a trail leading from the yard and heading in the direction of the Twelve Mile Creek, which is only a mile or two south. I inquired of the young minister if the creek was passable at that point, and when he assured me that it was I headed the team for it. It was in the month of November.

When I reached the bank of the creek, I had a moment of doubt. There was quite a lot of ice on it. But finally I put my trusty ponies into it. They got about halfway across. The creek being at high water and quite wide then, they found they could not get their front feet on top of the ice to break it, so they floundered and presently one of them lost his footing and went down in the water. That called for prompt action.

I was wearing a heavy, dog skin fur coat which I packed on the seat of the buggy and stepped into the water up to my waist and, of course, as cold as ice. I succeeded in getting the horse on its feet, broke the ice in front of the team, and then climbed back in the buggy. I reached the south bank of the creek, but I was thoroughly soaked up to my armpits and in

The Davies and Aveline Sproule family, LaFleche homestead, 1913.
Florence is second from the right. Photo in possession of the author.

a serious mess in freezing weather. I looked over the country ahead of
me, watching for smoke, and spotted one a little to the southwest of me.
I applied the whip to the team and headed for it at a gallop.

When I knocked at the door, it was opened by a tall man, Chesley E.
Sproule, who took the situation in at a glance. He reached behind him
and got a suit of long, dry underwear which he placed in my hands and
told me to forget about the team. He would take care of them.

He had a stove going full blast in the shack, and I promptly stripped
naked and proceeded to thaw myself out. On Chesley's return from the
barn, we had lunch together. Later in the afternoon, for it took some time
to dry all my clothes out, he suggested that we take a walk south a mile
and a half and meet his parents. We did so, and in his parents' home
that afternoon I met for the first time my wife, Florence Sproule. We had
dinner in the Sproule home, and I spent the night with Chesley, taking
the road next morning all dry and comfortable again.

Davies and Aveline Sproule had pioneered a thin-soil bush farm at
Mapleton, Nova Scotia, where they raised a family of seven children, four

boys and three girls, of whom Florence was the youngest. After Davies travelled west with a harvest excursion and discovered the rich soils of the prairies, the entire Sproule family emigrated to Saskatchewan, taking up homesteads ten miles south of LaFleche. The elder Sproules were almost sixty years of age when they began their second career of pioneer farming.

Wedding portrait of Charles and Florence, August 16, 1916. Photo taken in Regina on return from honeymoon in Qu'Appelle valley.

Florence was at that time a teacher at Harwood School on Section 15-7-5 W3rd. I was driving the whole country, so that it was not much trouble for me to show up at Harwood School in the late afternoon of the occasional Friday. This enabled me to maintain contact with the girl I wanted for my future wife, and the acquaintance grew and ripened until a year or so later I felt justified in making a proposal to her. The proposal was not immediately accepted, but I had lots of time and continued the suit with persistence, and finally in the goodness of God the deal was consummated. It was a deal which I, at least, have never had a moment's right to question. No man could possibly have picked out a better wife to share his lot in a new country.

Charles and Florence were married on August 1, 1916, in her parents' very modest home on their homestead. The newlyweds honeymooned in the Qu'Appelle Valley and became so fond of the Calling Lakes that later they kept a cottage there for many years. They took up residence at Limerick.

Later in 1916, his passion for politics drew Charles into another election and a caper that delivered a humbling lesson.

ENDNOTES

1. Bill Barry, *Geographic Names of Saskatchewan* (Regina: Peoples Places Publishing, 1998), 245.
2. William Allen, cited in George Britnell, *The Wheat Economy* (Toronto: University of Toronto Press, 1939), 79.
3. Peel, "R.M. 45," 207, 209.

CHAPTER 7
=====

1918 FLU EPIDEMIC

I n 1916, Limerick was part of the Moose Jaw County provincial constituency, represented by Liberal J.A. Sheppard, speaker of the Saskatchewan legislature. Sheppard, implicated in a bribery scandal involving a number of Liberal MLAS, resigned to seek vindication from his constituents. A by-election was called for December 5, 1916. It was a two-way race, with Sheppard running as an Independent Liberal against J.A. Chisholm, Conservative.

The evening before the election, my friend Dr. Herbert Gordon [Limerick veterinarian] and I sat down in my office to study the voters list. We were shocked at the extent to which it was loaded against us. I still recall it contained 160 names, and we estimated we would lose by 100 votes. Then Dr. Gordon thought up an incident which he had seen happen in a poll in Ontario and thought we might apply its principles here.

The weather was extremely cold, and it was a safe bet that the great bulk of the electors would present themselves in the afternoon. We had here at Limerick a large Luxembourg colony who were loyal to the Liberal Party, and they all had women at their side, women who spoke no English. Dr. Gordon suggested that we get up at five o'clock in the morning and set up an organization to collect all these ladies at the office and hold them without voting until the afternoon. Then, we surmised, if in the afternoon we exercised our right of swearing in every elector, there would be a large part of the voters list left without time to vote.

This was done in detail, and we packed the front room of the office [voting was held in the back of Charles' small office building on Main Street] with these lady electors who spoke maybe French or German but very little English. We did not vote any of them in the forenoon, but about two o'clock we started to bring them in to vote. When it came to swearing them, the deputy returning officer discovered that there was no French interpreter available, and so quite a few valuable moments went down the drain before we were able to locate a person who could speak French.

The voting went very slowly indeed, and outside the office there piled up an awful lot of Tory voters who were unable to get in and who became very restive, not to say hostile, as the moments slipped away. Finally, five o'clock arrived, and the poll was closed. When it was counted, we discovered to our delight that we had lost it by a mere eight votes instead of the 100 we had been anticipating.

However, it was not all that good. There was a terrific local outcry about the incident, none of which, mark you, was illegal, but I made up my mind that never again would I be a party to similar proceedings.

The Saskatchewan Election Act was later amended to prevent such machinations by providing that any voter present at a polling place at the close of the time for voting would be permitted to vote. And the caper was in vain. In spite of the contribution made at the Limerick poll by Charles and Dr. Gordon, Sheppard did not secure vindication but lost the election to the Conservative Chisholm, 2,058 to 2,148.

Two other by-elections were held in the fall of 1916, in the ridings of Regina and Kinistino, the latter made necessary by the resignation of another disgraced Liberal MLA. Both were won by acclamation by newly recruited Liberals, William Martin (see p. 228) and Charles Dunning (see p. 226). Martin resigned his seat in Parliament to succeed Premier Walter Scott, who had resigned, his failing health having rendered him incapable of dealing with the scandal facing his government.

Dunning, an English homesteader with a talent for management, had been the very successful general manager of the Saskatchewan Co-operative Elevator Company, a farmer-owned enterprise. Under

Dunning, the scec had grown from its beginning in 1911 to the operation of 230 country elevators by 1916 and was the largest grain-handling company in the world. Dunning was then recruited to Premier Martin's new Liberal government and appointed provincial treasurer.

A.J. Hindle, Charles' landlord in 1912, developed a yen for a political life and enlisted his former boarder to assist him in securing the Liberal nomination for the constituency of Willowbunch. The nomination was won, and so was the riding in the election that Premier Martin called for June 26, 1917. Hindle would hold the Willowbunch constituency in two more elections but then resign immediately following the election of June 2, 1925, to make way for a defeated cabinet minister.

Chisholm represented the constituency of Moose Jaw County for only six months. In that June 26, 1917, election, Dunning moved from Kinistino to Moose Jaw County, where he was easily elected in a province-wide Liberal landslide. The scandal forgotten, Martin's government secured fifty-one of the fifty-nine seats.

In Limerick, in his new constituency, Dunning found Charles, an active and effective Liberal supporter. The two developed a life-long friendship that continued as Dunning went on to become the premier of Saskatchewan and then minister of finance in the government of Prime Minister Mackenzie King.

In 1916 there occurred another event significant to the life of Charles Wilson. Although he had accepted his adopted country without reserve or condition and became a passionate Canadian, Charles retained his love for Ireland. And he was an ardent Irish nationalist who chafed at the British occupation of his homeland. Thus, with his heart, Charles cheered a group of fierce Irish patriots who rose against the British in Dublin during Easter 1916. The rebellion was short lived and failed, but it fired the first shots in what became the Irish War of Independence. The Troubles, as the conflict came to be called, would later reach all the way to Limerick, Saskatchewan.

The so-called Spanish influenza epidemic of 1918 that caused an estimated 50 million or more deaths throughout the world did not spare Saskatchewan or Limerick. Charles became deeply involved in the community's response.

For some unaccountable reason, the flu descended upon this community suddenly and like a black fog. Immediately, the school board and the village council met to consider the situation in which a good one-half of our people were confined to bed with the disease.

We had at that time a five-room schoolhouse, and they closed the school and converted it into a hospital. There was in town a hotel which was furnished but had never operated. We denuded the hotel of beds and pretty soon had the hospital crudely equipped but ready to function. From the moment we opened its doors, we had plenty of business.

We were fortunate in having in the community three registered nurses who volunteered to take charge of the hospital. Our doctor was very ill with the flu and unable to function at the time. These three ladies divided their day into three eight-hour shifts and took most competent charge of the situation. For staff, we went out to the country and closed the country schools and brought the teachers in as nurses' aides. We never met with a refusal or a dissent. I remember personally calling at Valance School, south of Melaval, and telling Miss Murphy, the teacher, that we needed her in Limerick. Without even consulting her school board over the phone, she dismissed the kids that afternoon and piled into my car and joined the staff at Limerick.

Actually we found in a day or two that we were running quite an excellent little hospital, every bed in which was full. In fact, we were continually adding more beds to it.

The overseer of the village came in to see me one afternoon and told me they wanted me to act as superintendent. Fortunately I had escaped the flu.

I told him, "You have a committee. The reeve of the municipality is a member of it. He is not a friend of mine, and I don't think I could work harmoniously with him. If you will call a meeting, and the meeting appoints me, I am willing to serve."

He came back in an hour and told me, "The situation is so serious here that the two of us have to be the meeting. There is nobody else in shape to come to it."

Under those circumstances, I waived my scruples and agreed to serve, and it would be impossible for anyone to receive from a community more wholehearted cooperation than I received from this one.

Everybody in the community laid down the tools of their job and placed themselves at our service. Since the disease was scattered all over the countryside, we had quite a problem to keep track of the many households. I can remember Nicholas Surdia, of Assiniboia, buying a new Chev car, pasting a big red cross on the windshield, and coming up and telling us that he was at our service. Since I was superintendent at the hospital, and since he had a new car, I told him I was going to give him all the territory from the Twelve Mile Lake to the Montana border. He was to drive it diligently every day, and, if he noticed a house that was not emitting smoke from the chimney, he was to call there and see what the situation was. It was amazing in how many cases he found that everybody in the house was on their back with the flu. He loaded them in the car and brought them to the hospital at Limerick. North of town, my old friend Davey Birss went up on a hilltop with a pair of binoculars and surveyed the flat land about him. If there was smoke coming out of a chimney, he did not worry. In the absence of smoke, he saw to it that someone called at that house in the next couple of hours. And so we kept pretty good track of our countryside and avoided any tragedies at least.

At the time, there was a heavy poker game in existence in the basement of the hotel organized by one Red Kilgore. Even Red we had on our staff. Every evening about midnight we would shuffle the beds of the hospital so as to bring those who were likely to pass away during the night convenient to the door so that they would be less of a nuisance to the other inmates when the event happened.

When the event did happen, the hospital sent a messenger down to advise Red Kilgore. His gang would promptly turn their cards face down on the table, pick up a stretcher which was in the corner of the room, come up to the hospital, load the deceased on the stretcher, and convey him to a little building owned by the village on Main Street which we used as a mortuary. The building was the village gaol.

We had a Jewish storekeeper in town, Louis Levitsky. I met him crossing the street one morning when we had five casualties in the morgue, and I told him, "Louis, I've got to have you to help bury these people."

He backed away from me and said, "The Jewish religion does not permit me to touch a dead gentile."

I respected that and went on into the office. Ten minutes later Louis came in and said, "To hell with the rules. I'm at your service."

I think quite a little nobility is involved in that truth.

My neighbour Archie Card lost his wife in the epidemic. The morning after the funeral, Archie turned up at the hospital with his car and said, "Here I am. Use me."

I had not thought of calling on Archie in the circumstances, and I think his action was one of nobility also.

We had the utmost difficulty in burying our dead. Albert Wright worked hard in the lumber yard for quite some time turning out plain, rough coffins, but he was unable to keep up with the demand, and a good number of our casualties were interred without any coffin. There was no other course of action open to us.

A hospital cannot be carried on without a kitchen. To meet that end, we brought in George Lundblom's threshing outfit cook car and installed it alongside the school. Two elderly ladies in the community, Mrs. Dean and Mrs. Boyk, volunteered for the kitchen. I used to see them every morning passing my house at a quarter to six on their way to work.

I went in one day to see how they were getting along and found them turning out such food as I venture to say was never served in a hospital before or since, and one could be forgiven for coming back across the line to eat the food.

Mrs. J.D. MacMillan had a large home in town which she vacated in order to allow us to use it as a dining room for our nursing staff. There was out in the country a crotchety old lady, Mrs. Mary Hunter, who immediately came in, donned a white cap and an apron, and took charge of the home. There at any time during the twenty-four hours excellent food was available to all our nurses.

One day I informed Mrs. Hunter that we had placed her on the payroll at $3.50 a day. She immediately scolded me and refused to continue on the payroll. She informed me that she came into the village to do her duty, and she did not propose to allow us to pay her.

People were deadly scared of the flu and of coming into contact with it or anyone who had it.

Most went around for weeks with scarves or masks over their noses and mouths. Nonetheless, this did not prevent many of those who were genuinely scared from coming into the hospital to make a donation of half a dozen dressed chickens, or a big chunk of beef, or vegetables.

A laundry was equally imperative. We hauled in another cook car, and two young Romanian boys who were in uniform, Curly Montan and Gus Cojocar, went to work in there. We bought them a gasoline-powered washing machine, and they had plenty of work. Every morning they washed all the soiled linens in the machine and hung them out to dry. Then they put them through a mangle, and they went back to the hospital, scrupulously clean.

If I needed a car to go out to the countryside, I was given the privilege to take possession of any car on the street, and I exercised this privilege freely and without demur from anyone, although, undoubtedly, I caused several people quite a lot of inconvenience.

The situation out in the rural area was at times tragic. A farmer phoned me one evening telling me to send someone out to his neighbour's farm, where they were all confined to bed with the flu. Their baby had died and was stacked on a pile of wood in the outside porch. We sent a car out and brought the rest of the family in to the hospital.

There were many bachelor farmers and isolated families in the rural community at that time. So far as I can recall, not one of them died alone. Our daily patrols were adequate to the point of assuring that. All the deaths that took place occurred at the hospital.

On the school staff was one Miss Wilson, a little, slender teacher who had come to us from Halifax. She had been there in 1917 when a huge explosion had taken place on a munitions ship in the harbour, causing a tremendous tragedy and many wounded and dead sailors and townsfolk. Miss Wilson had nursed through that great loss before she came here, and here she barged right into the flu epidemic and went to work on our staff.

She and a companion were sleeping above my office. One day I noticed through the office window the two of them coming off duty at four o'clock and going to their quarters. I followed them upstairs and told them I was happy to inform them that the load at the hospital had become greatly reduced, and we would not require them to go back on duty.

I went back downstairs, and by the time I arrived there I heard a commotion upstairs which was quite novel to me. I went up to see what the trouble was and found Miss Wilson in the throes of a very serious

61

attack of hysterics. I am convinced that, if I had just told her that she would have to serve for a month longer, she would have met the challenge without faltering, but relaxation came too quickly. She was a great and brave little lady.

We carried on the hospital for quite a long period, charging our patients five dollars a day, if they could afford it. Miss Allonby, my secretary, kept the books. In late October, the siege lifted, the hospital slowly emptied, and we were able to turn it back to the school board. The total amount we had spent was $13,000. Miss Allonby had kept track of the extent to which the hospital was used by residents of the village and residents of the rural municipality, and she divided the inevitable deficit, using these figures.

I went to the next meeting of the village council, and the next meeting of the RM council, and without hesitation each gave me a cheque to cover their share of the deficit. So we were able to close our doors without owing a dollar to anyone and thanking God that the number of deaths was not nearly so high as was at one time anticipated. All the same, we lost thirteen people during the epidemic.

I have seen this community meet many difficult situations with compassion and dignity, but I still believe that 1918 was Limerick's finest hour. During the continuance of that epidemic, we never wrote a letter to Regina requesting assistance from anyone, not from the Red Cross or the government. The community faced up to the challenge and discharged it by their own organizational efforts. I doubt if you could find in the Province of Saskatchewan another community which met such a challenge and relied entirely on their own resources without appealing to anyone for a dollar of assistance or assistance of any nature.

The flu pandemic weakened in 1919 and had run its course by 1920. By then, it had taken some 50,000 lives in Canada, 5,000 in Saskatchewan.

CHAPTER 8

POST-WAR PROSPERITY *and* DISTRESS

Tom Wilson came home from the war, apparently hale and hearty, though Charles detected a slight change in personality, a natural effect after two years on the front lines of that dreadful conflict. Tom re-entered the office of Wilson Brothers as if he had never left and set about business with gusto and enthusiasm.

Two of Florence's brothers, Chesley and Leroy, also served overseas with the Canadian Army. Both survived the carnage, but Leroy fell to the Spanish flu epidemic in England. Chesley came home unscathed but soon found his life on his homestead too dull and, with some assistance from Charles, set up business in farm loans and insurance in Rockglen, where he remained all his life.

Charles marvelled at Tom's ease with people. Tom was instantly liked by almost all. As his brother described it with his Irish turn of phrase, *Tom strode straight into the hearts of men.*

On one occasion, one of their clients, a farmer, told Charles,

"That brother of yours is crazy."
Surprised, I asked, "What do you mean?"
"I mean that he drove into my yard at two o'clock in the morning, got me out of bed, and made me pay my note."
And this was said without any trace of resentment or rancour.

Main Street, Limerick, 1923. Photo in posession of the author.

It was a prosperous time for Limerick and for all the towns and villages that had sprouted up along the CP rail line from Assiniboia to Shaunavon. Communities with grain elevators were placed eight to ten miles apart along the rail line to facilitate grain deliveries from farms. Grain, usually wheat, was then hauled by horse and wagon, and elevator placement was intended to ensure that a farmer could make a round trip in one day.

But there was no railway between Limerick and the American border, some sixty miles south. Thus, the communities along the eastern section of the Assiniboia–Shaunavon line served an unusually large territory until the line to Mankota was constructed in 1928. Several of the towns had more elevators than normal: Limerick had six; farther west, LaFleche had seven, and Kincaid boasted eight.

During its heyday, Limerick's business district boasted four livery barns, two lumber yards, three garages, two blacksmith shops, four restaurants, a movie theatre, five general stores, four farm implement agencies, at least two banks, a shoemaker, a harness shop, a drugstore, a bakery, two butcher shops, a tailor, a hardware store, a jeweller, a bowling alley, a Chinese laundry, a flour mill, three real estate and insurance offices, a pool room and barber shop, a newspaper (*Limerick News*), a medical office, the CPR telegraph and freight office, a telephone exchange, and a dray delivery. Also available were two midwives, a nurse, and the services of a number of tradespeople, including a stone mason, a plasterer, and a painter and decorator.

Homesteaders who located on those southern lands faced a serious hardship in getting their produce to market. Hauling grain took

place between October and March, after harvest and before spring made the roads impassable. A large crop meant a full winter's work, for horse-drawn grain wagons held only eighty bushels, or the larger tank box 120 bushels, requiring a four-horse team. When the snow became too deep for wheels, sleigh runners were placed under the grain wagons. Historian Bruce Peel described a farmer's trip to town:

> In preparation for the journey, the farmer loaded the wheat on the wagon the night before. Standing in the granary, ankle-deep in the wheat, he scooped up shovelful after shovelful and, with a rhythmic swing, tossed it into the wagon. Poets have waxed eloquent over the "golden grain," but no poet ever shovelled the stuff. Dust filled the air, irritated the shoveler's nose and throat, and itched his sweaty cheeks. Perspiration trickled a smeary path down his dusty cheeks. He puffed with exertion. His back ached from stooping and straightening as he shoveled the innumerable scoops needed to fill a wagon. Sooner or later, when he straightened his back to rest it, he caught it on one of the taut strands of twisted wire used to brace the granary. With dogged determination he stooped again to shovel. The wagon had to be loaded.
>
> Between two and five o'clock in the morning the farmer arose. Lantern in hand, he went to the stable to feed and harness his horses, while his wife prepared breakfast and packed a lunch. The half-frozen lunch would be eaten when the farmer stopped along the trail to feed his horses.

The four horses were hitched to the wagon. To facilitate the loading of the wagon, little hollows had been dug into which the hind wheels had been lowered. Now the horses tugged and strained to pull the wagon out of the hollows. The silence of the morning was broken by the shouts of the farmer urging his horses on, by the crack of his whip, by the metallic clanking of whipple trees and traces, and then by the crunching of wagon wheels on frozen ground.

In winter the farmer dressed in the warmest of woolen underwear, two pairs of socks, and two pairs of mitts. He wore sweaters and a heavy mackinaw. If the wind was extremely cold he walked beside or behind the wagon to keep his feet warm. And swung his arms around his body to keep them warm. The weight of his clothes was fatiguing.

On reaching town the grain was taken to one of the several elevators, and the horses then driven to the feed barn. Before the introduction of the automobile and truck, the feed barn was one of the most profitable businesses in a town. After dinner the farmer bought groceries and other supplies to take home, and started on the homeward trail late in the afternoon. If he lived in the southern part, he arrived in town too late in the day to return that night.

The waiting farm woman frequently walked outside in the dark to the corner of the shack. She strained her ears to catch the rattle of her husband's wagon. When the farmer reached home at ten or eleven o'clock at night, he had chores to do, his horses to unharness, and his supper to eat. Sometimes he loaded his wagon for a trip the following morning.

Finally, late at night the farmer dropped into bed to sleep the dreamless sleep of extreme fatigue; fatigue produced by the cold, the hunger, and the aching muscles he had experienced during the day. With physical exhaustion should come a feeling of contentment, but as his mind lost consciousness he thought of the morrow, or the day after, when he would have the same experience over again.[1]

Hauling grain from forty or fifty miles south of the railway meant a trip of three or four days, one way, and was expensive, exhausting, and often dangerous in winter. Plus, someone would have to be left in charge of the homestead to care for the remaining livestock. Little wonder that some homesteaders abandoned their farms. Thirty-one gave up and moved out of the area along the border south of the later town of Rockglen.[2]

The position of the railways was made clear in July 1920 when the CPR representative stated to the Better Farming Conference in Swift Current that "Future railway development in the southwest would depend on the ability of the farmers to demonstrate that they could farm profitably."[3]

The demand for better railway service became a hot political issue. In the spring of 1924, N.R. McTaggart, Progressive member of Parliament for the constituency of Maple Creek that shared with the riding of Assiniboia the entire southern portion of Saskatchewan, drove fifty-two miles south from Limerick to attend a railway meeting at Lonesome Butte. At the conclusion of the meeting, McTaggart learned that, to give him an understanding of the distances and difficulties that they endured hauling out their grain, the local farmers had intended to make the MP walk every mile back to town.[4]

When the Saskatchewan Wheat Pool was organized in 1924, its first facility was a grain dump purchased in Scobey, Montana, fourteen miles south of the American border, purchased to serve the farmers of the area north of the border who had no nearby Canadian railway until the CPR arrived in 1927 and the communities of Coronach and Rockglen were organized.[5] The Scobey facility provided little relief to the settlers at Lonesome Butte, who found it no closer than Limerick to the north.

In 1920, Tom Wilson married a woman he had met in Ireland while overseas. Georgie Bennett had lost all members of her family and accepted Tom's proposal to come to Canada. The successful manager of a hotel at Killarney, she was, Charles said, *gorgeous and brilliant.* But Georgie was a devout Roman Catholic, and Tom was a Protestant, and therein lay the seeds of difficulty. The marriage soon began to founder.

By 1921, the grievous mistake of allowing indiscriminate settle-ment in the former rangelands of southern Alberta and southwestern Saskatchewan had turned into tragedy. Crops failed almost totally in nearly every year after 1917 as precipitation from Calgary to Swift Current dropped well below normal and summer temperatures were scorching. In southern Alberta, the average crop production in 1917 fell to 10.7 bushels per acre, in 1918 to 4.9 bushels, and in 1919 to 1.4 bushels.[6] Conditions were as bad in the southwestern corner of Saskatchewan. South of the Cypress Hills and west of Robsart, an area of about 10,000 acres, the situation was extreme. In the six years 1914 to 1919, the annual precipitation averaged nine and a half inches, including the two very good years of 1915 and 1916. Excluding the two outstanding years, precipitation for the other four years was just over two and ahalf inches,[7] half that received by the Sahara Desert.

The light lands that should never have been broken began to blow away before the prevailing westerly winds. The agrarian agony was soon compounded by falling wheat prices. They stayed strong until 1920, then dropped by half. The word *relief* entered the western lexicon as municipal governments and the province initiated support programs in a vain effort to tide settlers over what was hoped would be a short-term calamity.

It was the second time in Saskatchewan's short history that public initiative had to be undertaken to alleviate widespread distress. In 1914, the advent of the First World War and an almost total crop failure in the southwestern portion of the province created serious hardship. Premier Scott's government began a road and bridge construction program to provide employment for drought-stricken farmers and set up a Debtor's Relief Bureau to mediate between farmers and their creditors with the intention of carrying the farmer through the crisis. The government resisted calls for a moratorium to protect farmers, fearing that such a drastic step would impair the credit of the entire province.[8]

Fortunately the bumper crop of 1915 ensured that the crisis was short lived. A wheat yield averaging just over twenty-five bushels per acre cured the problem.

In 1923, the federal government finally took some responsibility for the desperate situation in the former range country in the southwest and funded the removal of hundreds of settlers.[9] Thousands more moved out on their own. In a triangle lying west of Swift Current to the Alberta border, northwest to Estuary, and southwest to Govenlock, twenty-three municipalities lost 3,653 residents, or 12.3 percent of their population, from 1921 to 1926. Two of those municipalities lost a full third of their people. From 1917 to 1924, 10,469 farms in southwestern Saskatchewan were abandoned, a likely total of 30,000 people fleeing the region.[10] Across the Alberta border, the situation was more desperate, with 80 and 90 percent population losses.[11] Many trekked farther north, to the region beyond North Battleford, and to the Peace River Country. There they would find themselves even farther from railways than they had been in the undeveloped South Country.

In the belief that some of the failed crops were the result of inferior farming methods, the Saskatchewan government hosted a Better Farming Conference at Swift Current. Held over three days in early July 1920, the conference listened to farmers and agriculture academics, including some from the northern United States, and concluded by requesting the Saskatchewan government to institute a royal commission on the subject. The request was granted just one month later. On August 23, 1920, an order in council appointed W.J. Rutherford, dean of agriculture at the University of Saskatchewan, as chairman and John Bracken, president of the College of Agriculture, Winnipeg; George Spence, MLA and farmer of Monchy; Neil McTaggart,[12] farmer of Gull Lake; and H.O. Rowley, general manager of the Weyburn Security Bank, as members. The terms of reference somewhat understated the problem: "For some time past the condition of the farming industry in the western and southwestern portions of the province has not been very satisfactory."[13] The area of study was huge—west of the Soo Line running from Portal to Elbow and south of the Saskatchewan River.

The Better Farming Commission received advice from much the same group as the earlier conference, extending its reach to Montana, North Dakota, Minnesota, Wisconsin, and Illinois. Not surprisingly, no simple solutions were identified. The expansion of experimental stations, more study, and widespread education were recommended in the 1921 report, though much agricultural extension work had been carried out for several years. Every year since 1914 thousands of farmers had attended two- and three-day courses in dry farming techniques, and a Better Farming Train toured the province each summer, providing instruction and demonstrations to thousands more.[14]

Agrarian unrest led to political upheaval. Farmers' movements were moving toward direct electoral participation with surprising results. In October 1919, W.R. Motherwell, a highly regarded farm leader who had served as Saskatchewan minister of agriculture from 1905 to 1918, running as a Liberal in a by-election in the federal riding of Assiniboia, lost by over 5,000 votes to a United Farmer candidate. The next year saw the United Farmers of Ontario win a minority government and poorly organized farmer candidates reduce the Norris Liberal government in Manitoba to a minority. With political activism on the rise in rural Saskatchewan, Premier W.M. Martin manoeuvred adroitly to co-opt the policies and leaders of the farmers' movement and took the major step of dissassociating his Saskatchewan Liberal government from the federal Liberals of Mackenzie King.

In 1917, Charles, as secretary of the Maple Creek federal constituency, helped to arrange the unopposed nomination of J.A. Maharg as an Independent Unionist candidate. Maharg, whom Charles had first met on the Pole Trail in April 1912, had become a very prominent farm leader without a political label. He was easily elected in the 1917 general election.

Premier Martin induced Maharg to join his administration with the understanding that he could continue to support agrarian political activism. Martin called a provincial election for June 9, 1921. Maharg sat as MP for Maple Creek until the dissolution of Parliament but advised that he would not stand again and was elected by acclamation in the provincial seat of Morse as an Independent Pro-Government candidate. On June 14th, he was sworn in as Saskatchewan minister of agriculture.

Martin secured re-election of his Liberal government, but six candidates of the newly formed Progressive Party, the banner under which the farmer groups rallied, were elected.

A month later, in July, the United Farmers of Alberta entered the provincial election there and seized power, winning thirty-eight seats, all of them rural, out of sixty-one. By the next year, the Saskatchewan Liberal government would be the only survivor in four provinces.

A federal election was called for December 6, 1921. Martin and the members of his cabinet agreed that each could support the candidate of his choice without adherence to party labels.

Motherwell was again running as a federal candidate, this time in Regina. Martin spoke on his behalf at a meeting at City Hall on December 1ˢᵗ but went further and criticized policies of the Progressive Party.[15] The premier's speech was fully reported in the *Leader-Post* the next day.

On that Friday, Maharg, campaigning for the Progressive candidates, arrived in Limerick to address a meeting. After the meeting, he joined Charles at the Wilson home, where he had been invited to spend the night. There, for the first time, he read in the Regina *Leader-Post* a full account of the premier's speech.

In the morning, Maharg reported that he had spent a sleepless night. He accompanied Charles down to the office of Wilson Brothers, borrowed the telephone, called Premier Martin, and advised him that he would be resigning from cabinet. The resignation was formally effected on December 7ᵗʰ, the day after the federal election.

Maharg's resignation so highlighted what was seen as the premier's perfidy that it ended Martin's political career. Martin had "cooked his goose," as Professor David Smith described it.[16] Just four months later he was out of politics and appointed to the Court of Appeal.

In that federal election of December 1921, the government of Prime Minister Arthur Meighen was soundly defeated, and fifty-eight Progressive MPs were elected across Canada. In Saskatchewan, they took fifteen of the sixteen federal seats, Motherwell being the only Liberal elected. Alberta returned eight Progressives, two United Farmers, and two Labour MPs, shutting out the Liberals and Conservatives. Eleven of Manitoba's fifteen constituencies went to the Progressives. British Columbia elected three Progressive MPs, and the wave of protest ran as far east as Ontario, where twenty Progressives were elected, and even New Brunswick came up with one Progressive MP.

ENDNOTES

1. Peel, "R.M. 45," 250–52.
2. Canada, House of Commons, *Debates*, June 10, 1922.
3. Better Farming Conference, Swift Current, July 6–8, 1920, Saskatchewan Archives Board (hereafter SAB), R-261, 23-1.
4. *Debates*, March 13, 1924. 313.
5. Garry Lawrence Fairbairn, *From Prairie Roots: The Remarkable Story of Saskatchewan Wheat Pool* (Saskatoon: Western Producer Prairie Books, 1984), 68.
6. David C. Jones, *Empire of Dust* (Edmonton: University of Alberta Press, 1987), 108.
7. David Stenhouse, Consul, SK, to F. Hedley Auld, March 27, 1992, Deputy Minister of Agriculture Files, SAB, R-261, 23-1.
8. James William Brennan, "A Political History of Saskatchewan, 1905–1929" (PhD diss., University of Alberta, 1976), 216.
9. Bill Waiser, *Saskatchewan: A New History* (Calgary: Fifth House, 2005), 266.
10. Curtis R. McManus, *Happyland* (Calgary: University of Calgary Press, 2011), 70.
11. David C. Jones, ed., *We'll All Be Buried down Here: The Prairie Dryland Disaster 1917–1926* (Calgary: Alberta Historical Society, 1986), xli, xlii.
12. McTaggart was soon to be the MP for the constituency of Maple Creek. He was elected in 1921 as a Progressive and was defeated in 1925.
13. *Better Farming Conference.* Supra.
14. Waiser, *Saskatchewan*, 214.
15. David E. Smith, *Prairie Liberalism: The Liberal Party in Saskatchewan 1905–1971* (Toronto: University of Toronto Press, 1975), 91.
16. Ibid., 92.

CHAPTER 9

WILSON BROTHERS DISSOLVE

I n April 1922, Premier Martin resigned, and Charles Dunning succeeded him. Dunning continued the close relationship with the farm
interests that enabled the Liberal government in Saskatchewan to
survive the tempest in the agricultural sector. The government was so
willing to accommodate rural concerns that legislative requirements
were sometimes overlooked.

In 1921, the phrase "debt adjustment," a benign euphemism for debt
reduction, entered the lexicon of officialdom. In 1919, on behalf of the
Saskatchewan Association of Rural Municipalities, George Edwards
(see p. 224), a prominent western farm leader, proposed to the Saskatchewan government that some method was required to deal with
farm indebtedness. With the severe drop in farm commodity prices
following the war, Edwards claimed, farm debt had risen to a height
that was crippling the agricultural industry.[1]

Whether in response to Edwards' petition or not, in September
1921, the Saskatchewan government established the Debt Adjustment
Bureau within the Department of Agriculture. A debt adjustment
commissioner, Edward Oliver, was placed in charge of the new agency,
which came into being without any sanction or authority whatsoever
from the Legislative Assembly. No legislation was considered necessary.

The scheme required that a farmer seeking the bureau's assistance
turn over all the proceeds from his crops to the bureau for distribu-

tion among his creditors. Working entirely with the consent of the debtor and his creditors, without the threat of any compulsion other than the power of moratorium that the legislature had delegated to cabinet in the war year of 1914, the bureau had some success. By May 1, 1922, it had dealt with the applications of 3,500 farmers in financial difficulty, involving properties valued at $27 million.[2]

The Debt Adjustment Bureau, with its work varying according to the success of agricultural production, continued to operate even after 1926, when the delegation of the power of moratorium to cabinet came to an end. Its reach was extended by stationing inspectors in four judicial districts and appointing sheriffs to serve as representatives. In 1929, the bureau was replaced by a similar agency but with proper legislative authority. During the seven years of its existence, the bureau dealt with more than 30,000 cases of farmers in financial distress.

Officials of the governments of Alberta and Saskatchewan traded ideas and experience as both struggled to deal with their mutual problem of mounting farm debt. In 1922, the United Farmers of Alberta government, having come to power the year before, enacted the Drought Area Relief Act, providing for adjustment of debts with the power of moratorium. The next year that legislation was replaced by the Debt Adjustment Act, which extended the provisions across the province.

The need for some machinery to deal with the financial difficulties being experienced by the farm sector continued. In 1923, the price of wheat dropped to less than half its previous value. Farm operating costs remained high, and, in spite of the largest harvest since 1915, farmers were squeezed financially. They were very vocal in their protest. *The farmers were greatly upset, and I never heard more talk of violent action than I listened to that year.*

Charles was then supervising some 600 Victoria Trust and Savings Company mortgages scattered about Saskatchewan, payment became a problem, and he spoke with a large number of farmers over a wide area. He thought that the situation was dire and reported so when he attended the company's annual meeting in Lindsay. Charles' description of the economic woes in Saskatchewan was so moving that the president, William Flavelle, took him to Toronto to tell the story in some of the loan company head offices there, including the Canadian Bank of Commerce and National Trust. Then R.J. McLaughlin, KC, Victoria Trust's founder and vice president, who practised law in Toronto, took him home to dinner.

He was a legendary figure as far as I was concerned. He told me that he wanted me to accompany him to his den after dinner. He wanted to do a little talking to me.

When we were safely ensconced in front of a log fire in the den, he said, "Wilson, what I want to tell you is that you talk too much."

"Surely, Mr. McLaughlin, I have a grave situation to talk about, of which the creditors should be informed," I replied.

"No doubt. No doubt. I quite understand that. But, I repeat, I think you talk too much. I want you to listen to me. Your country is in the business of producing food, and eating is a habit that people never get away from. Fashions change and preferences change, but sitting down to the table three times a day is a universal habit. You are producing not merely food but fine food, excellent groceries, and, while I admit that you may have your ups and downs in some years, the long-term outlook for you is excellent, and everything will come out all right. I want you to remember that."

I remembered that through all the years of my active business career, and my thinking and my decisions were quite frequently coloured by the wise advice I received from Mr. McLaughlin.

The 1924 annual report of Victoria Trust contained a full account of the Saskatchewan investments to counter the impression that

Western securities have not been of the very best in the last few years. … Conditions have undoubtedly improved in the last two years, and as a result all signs point to a better and more hopeful feeling in the prairie provinces than has existed for some time. It has been said that the great majority of the farmers in Western Canada were in trouble such as they could not hope to escape from. There never has been any question but that the farmers of our West, with the exception of a relatively small percentage, would pull through any troubles that they might have if given reasonable time, and that without assistance from any one but as a result of their own industry and their own resourcefulness. The efforts and results of the last two years have proven this. I doubt if any other part of the world today offers as great an opportunity as the Western Provinces for the farmer of small means to obtain by his own energy and pluck a larger measure of success. Premier Dunning of Saskatchewan recently made reference in his budget speech to the financial situation of the Province of Saskatchewan in which he claimed that it was in better shape financially than any other province in the Dominion of Canada, with the single exception of Quebec.[3]

Saskatchewan ranked third in Canada only to Ontario and Quebec in population and members of Parliament and in economic prosperity. It was a condition that would be enjoyed only a few more years and then would vanish entirely, seemingly forever.

The comfortable situation at Limerick achieved at Wilson Brothers was not to last either. First the Irish War of Independence reached the two Irish brothers.

In 1922, Charles and Tom received an urgent telegram from Ireland. Their brother Nicholas, who operated a country store and small farm, had somehow brought on himself the wrath of the Irish Republican Army. He had been handed an ultimatum—be gone from Ireland within seventy-two hours, or suffer the consequences.

Neither Charles nor Tom was under any illusion about what those consequences would be. For more than two years, the IRA and the British Army had been waging the vicious conflict known as the Irish War of Independence, and life in Ireland had become cheap. The previous winter an Irish friend of Charles had returned to Ireland, planning an extended visit. When he was staying with a friend in Cork, the IRA had come in the middle of the night, taken the host out into the street, and shot him dead for the offence of "harbouring a stranger." The "stranger" returned to Canada on the next boat.

Charles and Tom put up the funds needed to enable their brother to flee Ireland. It was not a small amount. Nicholas was married and had four children.

Then came an event that dislocated the successful partnership between Charles and Tom.

In 1923, Premier Dunning, in Shaunavon to speak at an evening meeting, telephoned Charles at Limerick the next morning and asked him to join him on the CPR train when it reached Limerick and ride with him to Assiniboia. By then, Charles was a prominent Liberal in the South Country, and there was a matter the premier wanted to discuss. Charles did so. He could catch another train back to Limerick later in the afternoon.

"Charlie, you know my career. My success has lain in my ability to select the right men to work with me. I've been paying a lot of attention to that brother of yours. I want him. I want you to get him a seat, and I am telling you he won't remain six months as a private member."

I was very quiet and silent, thinking hard. I could have got him a seat. But I said, "Mr. Dunning, for the present, that's out. To make it clear to you, I've got to tell you his home is about to break up, and I'm unable to do anything to prevent it, and, until this situation resolves itself one way or the other, you know as well as I know—that's out."

An invitation to join Premier Dunning's government was a signal honour. The Saskatchewan cabinet then consisted of only six members other than the premier. But it was a political reality in the 1920s that divorce or separation was fatal to a public career. Charles spoke with Tom.

I stood there with my brother in my arms and told him, "Of all the young men in this province, you have the most brilliant prospects. There is only one obstacle. For God's sake, stop and think."

But the marriage did break up. Georgie left and moved to Montreal, where she re-entered the hotel industry and again achieved a management position in a major hotel.

Florence and I decided we were not going to take sides in the affair, but she clung to us as the only kin she had. And I stood in the huge cathedral in Montreal at her funeral as her only kin. Many people think that life is too short. You can sure pack a hell of a lot of experience and drama into one life.

The Village of Limerick was not big enough to contain this intense drama. Tom said, "You've got to buy me out, there's no other choice."

So I told him to go into the back office, spend a day there with what he knew to be the assets, and put a figure on it. He put a figure of $21,500 on it, brought it out to me in the evening.

I said, "Allow a little for over-evaluation—$20,000."

"Okay."

It was a huge price for a ten-year-old business in a community of just 325 people. It was the equivalent of nearly $275,000 in 2012.

It was like buying the national debt at the time. I was in a real jam, but I paid it. I came home and told this lady of mine I had made the deal, and I had no alternative to making this deal either, but I thought it was quite likely to put us in the poorhouse. But I asked if she would go along with me, and she did. And such is life—the deal made me a lot of money.

I went back to the office in the morning and told myself, "Charlie, if you are going to get over this, this office has got to make more money than it ever did. Where are you going to find it?"

And I picked on the hail insurance field. I knew I could get the volume of business and could do very well if I could get enough money to pay the companies cash and retain all the financing and commission for myself. I prepared a prospectus—Miss Allonby was with me then—and had her type out half a dozen copies. I said, "Now we'll start with one company. When they turn us down, we'll go to another. Before we get to the bottom of the list, maybe we'll click."

Most farmers considered hail insurance to be essential. But policies were sold on credit, the farmer signing a promissory note for the premium, payable, with a fairly high rate of interest, when the crop was harvested and sold. It was standard practice in the industry for the hail companies to carry notes. If there was a hail loss, of course the promissory note was deducted from the settlement. Charles proposed to carry those promissory notes himself and reap the interest as well as the commission on selling policies.

I sent the first prospectus to the United Grain Growers Security Company, asking for a loan of $20,000. I got a letter right back from Major Black, the manager. He said, "There'll be a directors' meeting in two weeks. I'll submit it. I think it may be possible."

And in two weeks he wrote me and said, "You have the money." The first one.

In the spring of 1924, Charles set about selling hail insurance over a large territory, so large that frequently he hit the road from Limerick on Monday at 3 a.m. Drawing on the loan granted by Major Black's company, he was so successful that he exceeded the originally approved amount and wound up borrowing $25,000 instead of $20,000.

That was a much better year for Charles. Although total production fell off somewhat, the wheat price recovered, and the yields were good in the Limerick region. All his hail notes were paid in the fall, the $25,000 was repaid, and Charles had won a substantial profit. Then Major Black came to visit, and Charles took him home to dinner.

He was playing with a couple of our children, he was a devil for having his pockets full of little magic things, and he looked across at me, and I was grinning.

"What are you grinning at, Charlie?" he asked.

"At you, sir," I replied.

"Why?"

"I was just thinking that you are the damndest businessman I ever had anything to do with. I applied to you last spring for a loan of $20,000. You didn't ask me for a property statement. Later on, I offered you security, and you wouldn't take it. You loaned me the money, and I ran an overdraft with you—it was $25,000—and you're paid. And I think it's a damned funny transaction."

He looked over at me quite seriously and said, "Not as funny as you think. Do you remember in the spring I told you that you would have to wait two weeks for a directors' meeting?"

"Yes," I said, "quite natural."

"Only there was no directors' meeting," he said. "I used that two weeks to make enquiries about Irish Charlie Wilson. And then I loaned you the money."

All my cockiness and conceit left me, and I was figuratively on my knees and thanking God.

Hail insurance policies are sold and placed in May and June, when the crops are sowed, the coverage identified, and the amount of protection decided on. The storm season is July and August. Except for the collection of the premium notes in the fall, it was a part-time industry for Charles but very lucrative.

Wilson Brothers (Charles continued the name after Tom left) became one of the three largest hail insurance local agents in Saskatchewan. Charles not only succeeded in paying out Tom but also was on his way to becoming a well-to-do man.

ENDNOTES

1. George F. Edwards, "Memoirs of George F. Edwards," SAB, S-A3, 1950, Accession Nos. 96–204–254–502, 28.
2. Report of the Debt Adjustment Commissioner, *Journals of the Legislative Assembly of the Province of Saskatchewan*, 1929, 186.
3. Victoria Trust and Savings Company, *Twenty-Ninth Annual Report* (Lindsay, ON: Victoria Trust and Savings Company, 1924), 12.

CHAPTER 10

PROSPEROUS DAYS

Premier Dunning and his Liberal government won solid re-election on June 2, 1925, with fifty out of sixty-three MLAS. The Progressive Party fielded forty candidates and secured 23 percent of the vote but elected only six MLAS, the same number as in 1921. Dunning won his seat of Moose Jaw County with over 71 percent of the vote.

Colonel James A. Cross, attorney general of Saskatchewan, unexpectedly lost his Regina seat in the 1925 election. The senior partner of the law firm Cross, Jonah, Hugg and Forbes, he was first elected to the legislature by Soldiers in Great Britain in the 1917 election, then won a seat in Regina in the 1921 election, and was included in the cabinet of Charles Dunning when he became premier on April 5, 1922.

Premier Dunning called on Charles, who induced his friend, Willowbunch MLA A.J. Hindle, to step aside for the good of the party. Hindle resigned, and a by-election was called for August 31, 1925. Cross was elected by acclamation and continued in his portfolio. He in turn would resign in 1927 to become a member of the Board of Transport Commissioners. He succeeded to the chairmanship in 1940. While in that position, almost twenty years after that 1925 crisis, Cross would remember his gratitude to Charles for his assistance in finding him a seat.

The year 1925 was even better in the western farming communities than 1924. Production rose substantially, and wheat prices remained strong. Charles' report to Victoria Trust and Savings Company was so upbeat that a portion was included in the annual report:

> The conditions in this Province as a whole have improved wonderfully during the past few years. Not only has the financial condition of the farmer improved in a manner not possible we think in any other country, but the general morale of the people is much better, and there is more contentment and more attachment to the land. The continued fair price for wheat has altered the outlook beyond recognition, and everyone now feels that farming is a business worth while and good enough for anybody. All this has developed a new demand for land, and in my locality farms cannot be bought or rented, so keen has the demand become.[1]

R.J. McLaughlin, KC, had been proven right, at least in the short term. Ten years later his philosophy would be severely tested in the longer term.

In the spring of 1926, Charles, by then prominent in the Saskatchewan insurance industry, was elected president of the Saskatchewan Insurance Agents Association, succeeding W.J. (Billy) Patterson of Windthorst. Patterson, a Liberal MLA since 1921, had resigned upon his appointment to cabinet, succeeding Premier Dunning as provincial treasurer. On March 7, 1928, during Charles' presidency, the association secured incorporation by private act of the Saskatchewan legislature.[2]

In 1925, Charles helped to organize and became a director of Central Canadian Insurance Company, with its head office in Winnipeg. All shareholders were insurance agents who directed business to the firm, a formula that was initially successful.

Early in 1926, Prime Minister Mackenzie King recruited Dunning to the federal cabinet. On February 26th, Dunning resigned as premier and was succeeded by James G. Gardiner (see pp.226–27).[3] Dunning was elected by acclamation in a by-election in Regina, moved to Ottawa, and immediately entered the King cabinet, initially as minister

Charles and Florence, with Kevin and Moira, and their Studebaker
in front of their recently renovated house, 1925. Photo in posession of the author.

of railways and canals. The King minority government fell, and Dunning spent a short spell in opposition until another election in 1926 restored the Liberals to office. Dunning became Canada's minister of finance in 1929. The close relationship between the minister and Irish Charlie at Limerick continued.

To Florence's delight, while visiting the Wilson home at Limerick, Premier Dunning suggested to Charles that he build a somewhat better residence. That awoke Charles to the realization that the modest bungalow that he and Florence had occupied since their marriage in 1916 had become a bit small for a family that then included two children, Kevin, born in 1917, and Moira, born in 1919. Two more children would follow, Sheila in 1926 and Garrett in 1932. The family crammed into two rooms while the rest of the existing house was demolished and a two-storey, five-bedroom residence built around them, completed in June 1925. With the automobile becoming an alternative to travel by train only, the Wilsons' new home soon became a magnet for travelling Liberal politicians. Forty years later Charles fondly recollected that every Liberal Saskatchewan premier, with the exception of the first, Walter Scott, had been a guest in their home, as well as any number of cabinet ministers, both provincial and federal, MPS, MLAS, and others who were prominent in public affairs.

In 1925, an old friend of Charles enlisted his support in securing the Liberal nomination for the federal constituency of Willow Bunch. Ten years earlier Charles was driving his team and buggy south of McCord when he came upon a man walking, carrying a small black bag. Charles picked up the pedestrian and thus met Dr. Thomas F. Donnelly, who practised at Kincaid. The doctor had been taken on an emergency call to the remote settlement of Horse Creek so suddenly that he had not had time to arrange a way to return home. From that beginning, a fast friendship developed, and Charles did not hesitate when Donnelly sought his help.

It was a tight battle for the nomination, partly because the Progressive movement was thought to be weakening and a Liberal nomination was considered to be a sure ticket to Ottawa in the upcoming election, and partly because J.G. Gardiner, minister of highways, was supporting his own candidate, Thomas Gamble, then the MLA for Bengough. There were several other candidates, including Thomas Gallant, a lawyer from Gravelbourg, a strong contender.

The Liberal delegates gathered in the Town Hall at LaFleche for the nominating convention. It ran until very late, with several ballots. Charles, perceiving that Gallant did not really want to be a member of Parliament but was developing credits for a judicial appointment, induced him to release a number of his delegates early to Donnelly to keep him in second place. The final ballot pushed Donnelly over the top.

Donnelly won the election on October 29, 1925, and was thereafter unbeatable, winning re-election in 1926, 1930, 1935, and 1940. He retired as a member of Parliament in 1945 and became a member of the Canadian Farm Loan Board, succeeding to the chairmanship in 1948, the year of his death.

During that first election campaign, a local minister approached Charles with a concern. There was a rumour that Donnelly had engaged in bootlegging. Charles said that he knew nothing about it but suggested that the minister address the question directly to Donnelly, who was speaking at a meeting in Limerick that evening.

"He'll give you a straight answer," Charles predicted. He was right.

"I bought a case of scotch legally before prohibition, and I sold it illegally after prohibition," the candidate frankly admitted to the open meeting.

That was the end of the matter. In Saskatchewan, prohibition had effectively ended in the spring of 1925.

Charles and Donnelly were the closest of friends. Unfortunately relations between Charles and James Gardiner were cool ever after, even though the two Liberals worked in the same political cause for many years. With some assistance from Charles, Thomas Gallant received his judicial appointment in 1930 and became the District Court judge at Gravelbourg.

His appointment was initially authorized by the Privy Council (the federal cabinet) on May 20, 1930. It remained to be confirmed by the governor general, but that had not taken place when an election was called for July 28, 1930, and routine business was suspended. The appointment remained in limbo.

By July, Charles instinctively believed that the election might not produce a Liberal victory. He got on the telephone to Ottawa, not a simple matter in those days, somehow reached Minister of Justice Ernest LaPointe, and raised the matter of Gallant's outstanding appointment. LaPointe immediately understood the problem and agreed to look after it.

Somehow he arranged that the order in council confirming the appointment was approved by the governor general on July 26th, two days before the election.[4] Defeated but still in office, Prime Minister Mackenzie King wrote to Gallant on July 31st advising him of his appointment.[5]

After the election, Charles took delight when the victorious Conservatives tried to pounce on what they thought was the still-vacant Gravelbourg appointment only to find that the door had been closed at the last possible moment.

Thomas Gallant served a long and distinguished career as the Gravelbourg District Court judge, retiring in 1951 at the age of seventy-four.

THE 1920S HAD BECOME SALAD DAYS, NOT ONLY FOR CHARLES and Florence Wilson but also for many in the South Country. Crops were good, wheat prices remained strong, and business was excellent along the eastern portion of the Assiniboia–Shaunavon CPR line that still served the huge territory south to the American border. Formal balls graced Limerick's town hall, with the ladies dressed in gowns imported from Winnipeg, often at a cost of several hundred dollars, a very large price. Automobiles were common, and travel became affordable.

Charles and Florence discovered Victoria and in 1927 spent the month of March at the CPR's Empress Hotel, then one of the grandest hotels of the British Empire. The great steamer trunk was brought up from the basement two weeks before departure, and Florence methodically stocked it with the clothing, including formal attire, that would be needed. Then, on the morning of leaving, Albert Schmidt, who operated the Limerick dray service, brought his wagon and matched pair of Belgians down the alley, and the steamer trunk was muscled aboard and delivered to the CPR station. When the train arrived in the early afternoon, the trunk would be loaded onto the baggage car together with the family's travelling luggage.

The family, Charles, Florence, and three children, ten-year-old Kevin, eight-year-old Moira, and Sheila, just a year old, clambered aboard the train destined for Moose Jaw, where they would arrive in time for supper (or dinner, as it came to be called) in the station dining room, complete with white linen tablecloths and the heavy silverware bearing the CPR crest. In the evening, they would be shown to their compartment in a sleeping car parked at the station. The car would be picked up and added to the transcontinental train when it arrived that night.

The plains of southern Alberta would be passing by their windows when the Wilson family awoke the next morning. Breakfast, and all meals, were taken in the dining car, again replete with starched linen and heavy silverware. Between meals, Florence was detailed to entertain the children while Charles, an inveterate cigarette smoker, adjourned to the lounge or parlour car at the rear of the train, reserved for first-class passengers. There his Irish charm quickly opened acquaintances with fellow passengers.

At Calgary, they enjoyed a station stop long enough to permit them to stretch their legs, and then they rolled through the glorious Rocky Mountains.

Arriving at Vancouver, the family transferred to one of the CPR's Princess vessels for the ferry run to Victoria, to the Inner Harbour dominated by the grand Empress Hotel. There they were accommodated in all the splendour that could be found in Canada in the late 1920s and rubbed shoulders with the business and political leaders of the day. Even so, the table d'hôte menu offered a four-course meal for $1.50. The Crystal Garden just across the street offered a saltwater swimming pool, ice cream, and activities a bit less formal than the Empress.

The Irish immigrant had come a long way from the elevator construction crews and the homestead at Dana.

The next year, 1928, was an outstanding one. Wheat production totalled just under a third of a billion bushels in Saskatchewan alone and well exceeded half a billion bushels in the three prairie provinces. Saskatchewan's total production of all grains hit a billion bushels for the first time. The price of wheat dropped somewhat but remained over a dollar a bushel, and with so many bushels on hand there were few complaints.

Some of the farmers in the Limerick district were so successful that Charles noted their production in the weather and travel diary that he had begun to maintain:

Ray Gravelle has 8,887 bushels on 1/2 section
Norman McIvor has 11,200 bushels on 1/2 section
E. Ballard has 6,610 of No. 1 wheat.

In the fall of 1928, J.B. Begg, a director of Victoria Trust and Savings Company, visited Saskatchewan, and Charles escorted him on a tour of inspection of the company's loans. Begg came away very impressed, as he reported to the annual meeting on February 5, 1929:

I was very much struck with the great improvement in the appearance of the country in the last eight or ten years. The original homesteaders, often misfit or amateur farmers, have either graduated into genuine farmers or given place to better men. Both in the matter of farm buildings and the cultivation of the land, it is quite apparent that the present owners know their business and have come to stay. They all seem to think there is no place like Saskatchewan for farming and are very enthusiastic, both about their Province and their business.

Begg made a favourable comment about Charles: "Our organization is as perfect there as it is in Ontario, and in our Manager, Mr. Wilson, we have a shrewd, competent and thoroughly trustworthy man, and as a collector one who cannot be beaten."

The introduction of mechanized farming in the west also impressed Begg, who had not seen a combine in operation before. He called it the combination, and those attending the annual meeting listened to a lengthy description of its function.

Begg's final comments provide insight to some generally held attitudes of the day. The annual report concludes the account of his address as follows:

> Referring to the foreigners against whom so much criticism has been raised, he said, "I want to take exception to a lot of this fool talk. To my notion some of these foreigners make the very best of settlers. They are willing to put up with the hardships and do the real spade work that belongs to the pioneer days, and more than that, they live on nothing and cook it themselves."[6]

ENDNOTES

1. Victoria Trust and Savings Company, *Thirtieth Annual Report* (Lindsay, ON: Victoria Trust and Savings Company, 1925), 10.
2. *Statutes of Saskatchewan*, 1928, c. 91.
3. On November 1, 1935, Gardiner followed Dunning's path, resigning as premier and joining Mackenzie King's cabinet in Ottawa. Patterson succeeded Gardiner as premier, holding that office until defeated in June 1944 by the CCF under T.C. Douglas.
4. LAC, PC-1098, 1930.
5. LAC, Mackenzie King Papers, J-1107-4-G, Mackenzie King to Thomas Gallant, July 31, 1930.
6. Victoria Trust and Savings Company, *Thirty-Third Annual Report* (Lindsay, ON: Victoria Trust and Savings Company, 1929), 17.

CHAPTER 11

DARK DAYS ARRIVE

J.B. Begg's comments about foreigners not only revealed a certain attitude on his part but also disclosed that while in Saskatchewan he had encountered an anti-alien fury that had taken voice. Perhaps he had failed to notice that anti-Catholicism had also entered the public discourse.

A new dimension had entered Saskatchewan politics in the form of the Ku Klux Klan, a slightly more benign but still virulent offshoot of the white-sheeted, Black-lynching night-riders of the American South. First surfacing in 1927, the Klan soon boasted an impressive membership of more than 20,000 and continued to grow even after the first organizers absconded with the membership funds. Its ambition to preserve Anglo-Saxon Protestant supremacy soon brought it into loose but effective association with the Orange Order, an anti-Catholic movement that originated in Ireland, emigrated to Ontario in 1830, and then moved west with Ontario homesteaders.

The Klan's bigoted rants were extensive. The editor of the *Maple Creek News* attended a Klan rally during the Maple Creek by-election held December 1, 1927, and then published a colourful description:

> Having no private source of information, *The News*, in common with the average man in the street believed that the pet

aversion of the K.K.K. was Niggers [*sic*], Chinks [*sic*], Jews and Catholics. After hearing the Klan speakers in The Grand last Thursday night, we now know that we must add Germans, Russians, Austrians, Frenchmen, Spaniards, Italians and Liberals. S'death, the list grows more respectable.

In 1928, both the Klan and the Orange Order turned to politics, openly supporting the Conservative Party in Saskatchewan. Their officers attended the Conservative convention in Saskatoon in March that chose Dr. J.T.M. Anderson as party leader. The convention adroitly managed to exclude any Catholic from election to any party office, including the fifty-seven-member executive. Murdoch MacPherson, a highly respected Regina lawyer, Conservative MLA, and soon-to-be Saskatchewan attorney general, carried out an investigation at the request of R.B. Bennett, leader of the Conservative Party of Canada. MacPherson reported that the Saskatchewan party was "purely Protestant and anti-Catholic."

Liberal premier James Gardiner attacked the Klan head-on. He issued a challenge to a public debate that was accepted by R.H. Hawkins, head organizer of the Klan. The debate was held in June in Gardiner's hometown of Lemberg before a standing-room-only crowd. It was a courageous act for the premier, but he failed to destroy the Klan as a political presence or an effective political enemy.

By 1929, storm clouds were gathering over Gardiner's Liberal government. Since the 1925 election, twelve by-elections had been held, and the Liberals had won them all, five by acclamation, though three of those were MLAs being elevated to cabinet where no contest was expected. But the last by-election, in Arm River on October 26, 1928, had been difficult, even ugly, as both the Klan and the Orange Order entered the contest. The Liberals squeaked through with just a fifty-nine-vote edge. A provincial election was called for June 6, 1929. The campaign was as vicious as that in Arm River and the results almost as close.

Toward the end of that 1929 election, in his home riding of Willow Bunch, Charles found that he had become the local issue. As usual, working indefatigably on behalf of the Liberal candidate, Charles

Johnson, Charles addressed a public meeting at Limerick on June 3rd, three days before the election. Speaking bluntly, in the style of the campaign, Charles took on the *Assiniboia Reporter,* a tabloid newspaper that was the local party organ of the Conservative Party and had been strong on the attack against the Liberals. Charles told his audience that the editor had been convicted of offences under the Liquor Act and had served time in jail.

The next day the editor commenced a lawsuit against Charles in the Assiniboia Court House, claiming slander and damages of $25,000. The lawsuit stated that Charles had said "There is a paper published in the Town adjoining us which is run by a man who is convicted of bootlegging and who has done 60 days. And where do you find this paper? Solidly behind Mr. Gibbons." Gibbons, of course, was the Conservative candidate.

On June 5th, the day before the election, the front page of the *Assiniboia Reporter* was devoted to the lawsuit under a page-wide, forty-eight-point headline: "$25,000 Suit Filed against Wilson: Slanderous Charges Are Made from Liberal Platform." Other stories were headed "Chas. Wilson, Limerick, Faces Slander Charge—Writ Issued" and "Life Long Liberal Workers Disgusted by Machine Tactics." The editor admitted to having been convicted of minor offences under the Liquor Act but did not claim, either in his paper or, strangely, in the lawsuit itself, that the statement supposedly made by Charles was untrue.

Charles, justly proud of his command of English, was more offended by the bad grammar attributed to him than the lawsuit, which he recognized as merely a last-ditch political ploy. In that he was right, for the writ was never served. But Charles did see the paper as a serious attack on his reputation and credibility, perhaps taking it far more seriously than he should have, and was concerned that the controversy might cost the Liberals the Willow Bunch seat.

The last Liberal meeting of the campaign was held on election eve at Flintoft, in the middle of the Romanian settlement. Charles shared the platform with the candidate, Charles Johnson, but requested that Johnson speak first, and only briefly, leaving most of the meeting to him.

When it came my turn to stand up, I spoke very intimately to my audience. I reminded them that a few years before I had come amongst them unknown and without an inheritance, and by their patronage they had built me a successful business and enabled me to establish a home. I then told them that, "If at this stage of the game you are through with me, and do not need me any longer, tomorrow is your opportunity because, if I lose this election, I will tell you very frankly that I cannot remain at Limerick. I will have to seek another location and start to build from the ground up again."

In the wind-up, I told them, "We will now sing 'God Save the King,' and after that I want all of you who are not going to be with us tomorrow to leave the hall. We still have two polls here to organize, and we want to turn this meeting into a Liberal conference."

They sang "God Save the King" I think more lustily than I have ever heard it sung, and then not one single person departed from the hall.

I stood up again and said, "Perhaps you did not understand me," and I explained the situation to them again and sat down, and still nobody left. Once again I stood up, and by this time my eyes were really shining.

I asked them, "Does this mean that you're all going to be with us tomorrow?"

There was applause and a roar from the audience. "That's what it means, Charlie."

I never felt more elated or flattered in my long life, because there must have been 400 people in that hall.

The next day, election day, the Flintoft poll voted strongly Liberal, fifty-three Liberal to nineteen Conservative, while the neighbouring poll, Gollier, was even stronger, ninety-four to twenty-two. The two polls contributed 106 votes to the Liberal overall plurality in Willow Bunch of just 107. The total vote was Liberal 4,423, Conservative 4,316.

Willow Bunch remained Liberal, but the Gardiner government lost its majority, winning just twenty-eight of the sixty-three seats, while the Conservatives came up with twenty-four seats. But also elected were six Independents and five Progressives, enough for a coalition to dislodge the Liberals. And so it was. Premier Gardiner remained in office until the legislature met on September 4th. Two days later, after

Headline in *The Assiniboia Reader*. Photo in posession of the author.

defeat on a clear vote of confidence, he resigned. Conservative leader J.T.M. Anderson became premier, at the head of what was called a "cooperative" government.

The elation of the Conservatives at finally achieving office would soon turn to despair. Six weeks after they assumed power, on October 29, 1929, stock markets collapsed and ushered in the worldwide Great Depression. In Saskatchewan, the agony to come would be compounded by the added calamities of a ten-year drought and falling farm commodity prices.

On June 18, 1929, Charles noted in his diary: *Rain yesterday quit about 7 p.m. A good rain ordinarily but of no account on land as dry as ours is now. Some parts of our district have more moisture, but on the whole our situation is quite critical. Only a very big rain can save it.* On July 27th: *Much late summer fallow does not seem to be developing any kernel. Blossom is dead, imprisoned in the chaff.*

In the fall of 1925, the CPR had commenced construction of a new line branching west from Maxstone, the second point south of Assiniboia on the Rockglen line, and three years later trains were in operation. A new string of communities sprang up: Stonehenge, Lakenheath, Flintoft, Wood Mountain, Fir Mountain, Glentworth, McCord, Ferland, and Mankota, the end of the line. Much of the trading area enjoyed by the towns along the Assiniboia–Kincaid portion of the Shaunavon line was sliced away, the first blow of a triple economic whammy that turned out the lights in many businesses. The second and third blows arrived two years later with the beginning of a decade of crop failures accompanied by a collapse in grain and livestock prices.

In 1930, the CPR extended another line, this time west from Rockglen twenty-three miles to Killdeer, just seven miles from Lonesome Butte, whose farmers finally acquired reasonable access to a grain elevator, but it was too late. The Killdeer line arrived just in time for the Dirty Thirties, and for years it was said that the line hauled in more bushels of seed wheat in the spring than it carried out after harvest.

The 1929 Saskatchewan wheat crop fell to half of what it had been in 1928, but net farm income fell even more disastrously, from $184,665,000 in 1928 to just $51,321,000, less than a third. At that, it was higher than it would be for another ten years and much higher than in five of those years when net farm income fell well into the negative range.

On December 31, 1929, Charles noted in his diary: *A little colder. Snowing a little and blowing. The welcome end of a darned anxious year.*

The 1929 annual meeting of the Victoria Trust and Savings Company, held on February 4, 1930, made moderate reference to problems with the Saskatchewan investments. The president, William Flavelle, told shareholders that, "In the West, we have our representative, Mr. Charles Wilson, of Limerick, a man of unusual ability, who is in close touch with every individual loan, and while he reports that in some districts borrowers will not be able to meet their obligations this year, yet on the whole the percentage of such will be relatively small."[1]

The managing director, C.E. Weeks, added that

> In Saskatchewan our interest collections last year were about $83,000.00, as against $94,000.00 for the previous year on practically the same investments, and this largely accounts for the slight shrinkage in our total interest collections for the year, but when it is remembered that the West only had half a crop last year I am sure you will agree that our borrowers in Saskatchewan gave a good account of themselves and proved themselves worthy of confidence and credit.[2]

Wilson Brothers office staff, 1931. L to R: Miss Maude Lindsay,
Mrs. Blanche Johnson, Charles Wilson, James Lawson, Miss Lizzie Allonby.
Photo in possession of the author.

Politically, things went from bad to worse for the Liberal Party
in the federal election of July 28, 1930. In June, Charles Dunning, by
then finance minister, was nominated in Regina by acclamation at
a huge rally in the armouries. According to the Regina *Leader-Post*,
more than 5,000 enthusiastic Liberals were in attendance. But on
election day, Dunning was defeated by more than 3,500 votes, and the
Conservatives under R.B. Bennett unseated Mackenzie King, winning
134 seats, a comfortable majority in the 245-seat House of Commons.
The Liberals fell to ninety seats, and the remaining twenty-one were
scattered among the United Farmers of Alberta and a few Progressives
and Independents. Saskatchewan was the only province other than
Quebec to elect more Liberal MPs than Conservatives, thirteen to six,
with two Progressives. Dunning returned to the business community
and was appointed president of Ontario Equitable Life and Accident
Insurance Company of Waterloo. Five years later he would again be
recruited to political life and join another Mackenzie King cabinet,
again as finance minister.

Charles struggled to keep Wilson Brothers afloat at Limerick. The
1930 provincial wheat yield recovered somewhat from 1929, but the

price fell by half, to an average elevator price of less than fifty cents a bushel. Net farm income dropped to $38,202. In the South Country, crop yields were not only well below the provincial average but almost a total failure, and economic misery took hold.

The Wilson Brothers office, with four employees, had a hefty payroll to meet. Lizzie Allonby was still serving as office manager, James Lawson handled the insurance files and assisted with the mortgage supervision, and there were two secretaries, Maude Lindsay and Blanche Johnston.

Charles was still operating the hail insurance business, suddenly no longer lucrative. At the request of Major Black, whose firm had gone on to other priorities, the credit line had been moved to the Bank of Montreal, where it ran to some $35,000 each season. In the fall of 1930, the great majority of the hail notes remained unpaid, leaving Charles to deal with the bank loan.

The hail notes were far from the only debt obligations that remained unpaid that fall. Farmers were so cash-strapped that creditors all over the west suddenly found that collections totally dried up. Banks, loan companies, implement dealers, merchants, what-have-you, secured or unsecured, accustomed to granting credit throughout the spring and summer with payment after harvest, found the second half of the equation almost completely missing. Rural municipalities saw tax collections evaporate.

At Limerick and other South Country communities, when it became obvious that the 1929 harvest would not produce enough to enable farmers to meet all their obligations, distressing scenes became common. Unsecured creditors, such as merchants, fearing that there would be nothing left for them behind secured creditors, banks, and loan and implement companies, hired trucks, drove right out to the threshing, and demanded grain on account almost as it came out of the separator spout.

One of the early casualties of the sudden downturn was the Saskatchewan Wheat Pool. Organized in 1924, it enjoyed early success. In 1925, it purchased the elevators of the Saskatchewan Co-Operative Elevators and in the next year handled 90 million bushels, the "largest quantity of grain ever handled by a single organization through its own facilities in any country in the world."[3] By 1928–29, it was operating 970 elevators and was a huge player in the grain purchase industry. Then it undertook an initiative to enter the marketing side. It had long been a dream among western farmers to take control of the sale of their own product.

The Saskatchewan Wheat Pool, with its counterparts in Alberta and Manitoba, formed a partnership as the Central Selling Agency and entered the parlous world of international export sales of wheat. In 1927, Ontario wheat was included. The Central Selling Agency, based on the premise of providing one price equally to all its members, was also initially successful. But it came to grief when it set the initial price for the 1929 wheat crop at a dollar a bushel, a decision that was considered conservative at the time and resisted demands for a higher level. The Winnipeg price then was $1.44 and soon rose to $1.78. But soon the price began to fall, and the agency compounded its exposure by buying more wheat and holding back its position, hoping to improve the price. Nothing worked as a worldwide wheat surplus took an inexorable toll on the price. By December 1930, the Winnipeg price dropped to fifty cents a bushel.

The final result was a huge loss to the Central Selling Agency of $22.1 million, of which $13.3 million was shouldered by the Saskatchewan Wheat Pool. The Saskatchewan government came to the rescue with a bond issue. The Wheat Pool undertook repayment in twenty equal annual amortized payments, at last retiring the debt in 1949.

In the spring of 1929, one of the last legislative initiatives of the Liberal government of Premier James Gardiner was introduction of the Debt Adjustment Act, which finally gave legitimacy to the Debt Adjustment Bureau, which had been operating for nearly eight years. The new legislation was operationalized by the cooperative government of Premier J.T.M. Anderson, and in 1930 Andrew J. Hosie (see p. 225), a prominent Regina businessman and strong Conservative, was appointed commissioner and later chairman.

Andrew and Charles were close business and personal friends in spite of their opposing political views. They were to collaborate unofficially as both sought solutions to the farm debt crisis.

The Anderson government amended the Debt Adjustment Act in every year that it held office, 1931, 1932, 1933, and 1934, each time increasing the authority of the board. In 1931, a limited power of moratorium was conferred, enabling the board to issue a certificate preventing legal proceedings against a named debtor. In 1933, the act provided that no action of any kind respecting the contractual obligations of a Saskatchewan resident could be taken without the permission of the board, a province-wide moratorium.

But even with the amendments, the Debt Adjustment Act contained no power to "adjust" debts, merely to postpone them. Any reduction of either principal or interest could be brought about only by an agreement between debtor and creditor.

In the South Country, particularly along the Assiniboia–Shaunavon CPR line, prosperity vanished, seemingly overnight, and was replaced by hardship more extreme than any yet experienced. Conditions were so serious that they attracted the attention of the Ontario press.

ENDNOTES

1. Victoria Trust and Savings Company, *Thirty-fourth Annual Report* (Lindsay, ON: Victoria Trust and Savings Company, 1929), 10.
2. Ibid., 14.
3. Fairbairn, *From Prairie Roots*, 67.

CHAPTER 12

DROUGHT

In January 1931, the *Toronto Star* sent journalist Wilfred Eggleston west
to investigate the accounts of economic suffering. He found his way
into the office of Wilson Brothers, made the acquaintance of Charles,
and filed his first story under the dateline "Limerick, Sask., Jan. 12, 1931"
and the heading "Drought, Tumbling Prices Ruining Western Farmers." Eggleston reported that not one farmer in the Rural Municipality
of Stonehenge, which surrounded Limerick, made expenses in 1930.

Why? Two reasons: A crop failure from drought, the second
in successive years. The tumbling of grain prices to a point ten,
twenty, thirty cents a bushel below the cost of production. …
Sixteen bushels per acre is the long-term average for the whole
prairie. The best districts, like Stonehenge, average eighteen to
twenty. I throw that in as introduction to the following:
Irish Charlie Wilson … gave me verse and chapter for the
following experiences, taken quite at random:
Farmer A, southeast of Limerick, 320 acres, threshed 700
bushels of wheat.
Farmer B, southwest of Limerick, 180 acres in crop, threshed
400 bushels.
Farmer C, southeast of Limerick, 180 acres in crop, threshed
540 bushels.

Farmer D, northeast of Limerick, 480 acres, threshed 700 bushels.

Farmer E, southwest of Limerick, farms 640 acres, threshed 980 bushels.

As Irish Charlie explained, these yields were very little more than would be required for seed next spring, and in no case would the total crop, ignoring the need for holding back seed, pay, at the low prices prevailing for wheat, the thresh bill and harvest labour. Obviously, nothing was left for taxes, interest, feed, oil, machinery repairs or groceries for the winter.[1]

Eggleston filed his next story, "Affluent Two Years Ago, Farmers Now Destitute," from Kincaid on January 14th:

Vast quantities of coal, flour and other groceries, clothing, feed and seed are being shipped into these villages and distributed as relief. An 800 pound bale of clothing collected by the Guelph, Ont. branch, women's missionary society of the United Church was distributed by Rev. John McKnight at LaFleche a few weeks ago. A second bale, weighing 1,250 pounds, from the ladies at Niagara Falls, followed shortly afterwards, and was soon parcelled out among the needy farmers. ... Here at Kincaid, four tons of clothing, provided by the W.M.S., the Red Cross and other relief organizations has been sent out already. One hundred and fifty families have been supplied with clothing by one pastor alone. ... Stonehenge distributed among its 690 farmers last winter $60,000 in hay, oats, flour and coal, and another $15,000 in seed grain—a per capita allowance of about $110. This winter feed is only about a third as expensive; even with that allowance, the gross expenditure will be greater.

Wood River rural municipality (a square of level prairie 18 miles each way, with LaFleche as its urban centre) has distributed this winter to date 600 tons of relief coal; is shipping in 100,000 bushels of oats and barley for feed, and has assisted 25 families with other groceries. ... In normal years Kincaid ships out 1,500,000 bushels of wheat and ships in only an occasional carload of feed oats. This season 5,000 bushels of wheat have gone out, and 95,000 of feed have been shipped in.

Eggleston's next story was filed from Hazenmore on January 15th under the heart-rending heading "Youth Dresses in Gunny Sack and Families Eat Gophers: Further Pitiful Stories of Privation Come from Canada's West." There was a touch of sensationalism in the heading. Eggleston admitted that he had been unable to verify the eating of gophers, and that he knew it to be an old prairie tale, but he did secure a first-hand account from Reverend Rondeau of the United Church, who in October had witnessed children wearing nothing but gunny sacks with holes for arms and legs. The pastor also stated that he knew personally of privation diets: "Bread and syrup, with a few potatoes for variety, is not uncommonly the whole menu for days at a time." Rondeau had heard of families trying to subsist on soup made only from Russian thistle.

"Starvation's Spectre Rises above Prairie Wheat Farms: 'I'll Starve with You,' Faithful Minister Tells His Parishioners"; "Villages Deserted." These were the headings over Eggleston's report from Shaunavon on January 16th. Again his story did not quite come up to the heading, but it did contain accurate accounts of hardship.

Tax collections are the despair of municipal officials. Here are the exact figures for Wood River rural municipality, which surrounds the town of LaFleche: In 1928 no less than 87 per cent of taxes levied that year were collected the same year, and for several years previous to that the average had been well over 80. In 1929, the first drought year, the percentage had fallen to 63 per cent. In 1930, when drought was combined with extremely low prices, it was only 36 per cent.

The towns and villages along the Assiniboia–Shaunavon CPR line were not deserted, but they were suffering a loss of population as many residents, particularly those running small businesses, moved away in search of better conditions. That population drain would continue, in time turning some of those communities into ghost towns.

Eggleston filed this story from Assiniboia on January 17th. He visited the local elevator and secured the day's grain prices.

No. 1 wheat, 35 cents a bushel; No. 2 wheat, 32 cents; No. 3 wheat, 28 cents.

Oats, highest grade, 15 cents a bushel; barley, highest grade, 9 cents; rye, highest grade, 12 cents; flax, 76 cents.

Dairy butter today is quoted at 10 cents a pound to the farmers here. Fresh eggs, 15 to 20 cents a dozen.

The *Toronto Star*'s reporter continued his tour, filing stories from Moose Jaw, Saskatoon, and then North Battleford before moving west to Edmonton and then down to Calgary and Lethbridge. At North Battleford, Eggleston found farmers who had harvested a bumper crop in 1930, thirty-five bushels of wheat per acre and 100 bushels of oats. But no one had made money, and most had not even cleared expenses. A wet fall had left everyone with tough grain, which knocked the prices down to the floor. One carload of 1,508 bushels of wheat that graded Number Three Tough produced a cheque of $277.69. A delivery of 700 bushels of oats brought only thirty-four dollars.

At Calgary, Eggleston attended the annual meeting of the United Farmers of Alberta, the governing party in that province. There he shared dinner with the UFA director from Peace River, an area that, like North Battleford, had enjoyed a bumper crop in 1930 but had also suffered a wet harvest. Eggleston's dinner companion provided a grim account of grain marketing in the Peace River area:

We're so far from our markets that the freight rate at the present time reduces the value of our grain almost to zero. Take wheat at today's prices. It was 53 cents a bushel, Winnipeg, for No. 1. Our average grade is No. 2 Tough, which is nine cents less, or 44 cents a bushel, Winnipeg. Deduct 21 cents for freight, leaving us 23 cents a bushel net at the elevator.

Now, follow that a step further. In our country we can't thresh for less than 12 cents a bushel. Deduct that and you have 11 cents to cover harvesting, hauling, twine, seed, seeding operations, plowing and cultivating, not to speak of taxes, hail insurance, depreciation of machinery, oil and gasoline.

I happen to be only a mile from town. But twenty per cent of our farmers live more than 20 miles from town. Motor trucks haul most of their grain. If it's done by contract, it costs a half a cent a mile. If a farmer, 22 miles out of Grande Prairie, sent No. 2 Tough to town today, he would get 12 cents a bushel after paying hauling, which would exactly pay threshing expenses, and leave him nothing to pay anything more.

As for the coarse grains, they're quite out of the question. It would be money out of pocket for most of our farmers to ship them. They can't afford to ship them, if you like to put it that way. ("Descent in Price of Grain Stuns Peace River Farmer," *Toronto Star*, January 28, 1931.)

Eggleston received confirmation of the dire situation facing coarse grains in the Peace River country from Premier John Brownlee, who joined a group chatting with the reporter in a hallway at the convention. The premier reported that "I had a letter yesterday from one of the grain companies informing me that at Grande Prairie, Wembley, Hythe and Dunsdale, they're paying one-half cent a bushel for oats. Think of it—two thousand bushels bring the farmer of that region the magnificent sum of ten dollars! Tell that to the readers of the *Toronto Star*." (January 30, 1931)

Eggleston had been alerted to the possibility that he might encounter serious talk among the desperate western farmers of seceding from eastern Canada. In Saskatoon, he found evidence of such desperation when he interviewed George H. Williams, then president of the 30,000-member United Farmers of Canada, Saskatchewan Section. (In 1938, Williams became leader of the CCF opposition in the Saskatchewan legislature.) "It is the financial control of the west by the east which we object to most," Williams told the reporter. "For years the profits of farming in western Canada have been sucked east through the banks, the mortgage and loan companies, farm implement companies and so forth, so that today 98 per cent of our farmers have no clear title to their land." Williams advocated what he called the "socialization of capital," which somehow would do away with interest rates. ("Prairie Political Pot Seethes, May Bubble Into Secession." *Toronto Star*, January 24, 1931)

A motion to secede from eastern Canada came to a vote at the UFA convention in Calgary. Eggleston reported that it was overwhelmingly defeated, achieving the support of no more than ten delegates out of 400 in attendance. (*Toronto Star*, January 29, 1931)

Premier Brownlee, fully aware that a radical element of the UFA was promoting Russian-style communism as a panacea, urged the delegates to study the subject and recommended a book that he had recently read. The premier received widespread applause when he pointed out that "Whatever is good about Russia you can have—by the exercise of your vote." (January 30, 1931) When Eggleston returned to Toronto, he wrote an opinion piece (February 5, 1931) for the *Toronto Star* entitled "The Western Wheat Grower and His State of Mind." He likened the farmers of the west to a wild animal caught in a trap, in its fear and rage biting anything and everything within reach, "and even inflicting wounds on any kindly-disposed observer who may approach with the object of setting it free."

> This explains, to my satisfaction at least, the prevalence of such extravagant talk as secession, repudiation of debts, wiping out of interest, boycott of stores, and Communism in various guises, which is heard in widely scattered areas of the west today. ... It is possible, I think, to be keenly aware of and profoundly sympathetic towards the western grain grower in his hours of bitter adversity while continuing to see and point out the fallacy of some of his solutions and the unreason of some of his criticisms.

In the west, as "the hours of bitter adversity" lengthened into seemingly never-ending years, and the fiscal nightmare became ever more horrifying, some reasonable and thoughtful leaders began to consider that there might be much less "fallacy" and "extravagance" in solutions such as the "repudiation of debts" and the "wiping out of interest." (See Eggleston, above.)

"The worst year in the history of Saskatchewan." That was the description given to 1931 by C.E. Weeks, managing director of the Victoria Trust and Savings Company, at the annual meeting in Lindsay, February 2, 1932.[2] No one out in Saskatchewan who had suffered through 1931 was likely to disagree with that assessment, at least not until even worse years came along, and these were not long in coming.

In many areas of the South Country, the crop was a total, utter, complete failure.

Our district has suffered from three crop failures. The crop of 1929 was a very short crop, the crop of 1930 still shorter, while 1931 amounted to almost nothing. In 1928, our farmers delivered to the local elevators around 850,000 bushels of wheat, while in 1931 this fell to 4,000 bushels.[3]

Those 4,000 bushels delivered at Limerick were 4,000 more than were delivered at three points on the new CPR Mankota line. The delivery in 1931 at McCord was zero, at Ferland zero, at Mankota zero. In 1928, their first year, those three delivery points handled, respectively, 383,000 bushels, 156,000 bushels, and 364,000 bushels.[4]

With the wheat price at the elevator still in the thirty cents a bushel range, it is little wonder that net farm income in Saskatchewan fell into the red as expenses outweighed the value of production by over $31 million.[5] Farm income would not hit the black again for four more desperate years—and then barely for two years before it again sank one more horrific time.

In July 1931, Florence Wilson drove a carload of young girls, her daughters Moira and Sheila among them, to a swimming hole in the Wood River between LaFleche and Gravelbourg. After a cooling frolic in the water, and a picnic lunch, they were driving home in the late afternoon when the sky suddenly darkened. Looking back westward, Florence saw that the sun had been blotted out by a towering, boiling black cloud rushing toward them at an incredible speed.

Astonished, and unaware of what was coming, her first thought was that the car would be unsafe. She shepherded her charges to a spot well behind the car and formed them into a close circle, arms around each other and heads bowed into the centre. Thus they endured their first dust storm, lashed by stinging gravel, their eyes, noses, and mouths filling with grit.

It was not the approved method of dealing with a dust storm, but as the storms became too-common occurrences Florence, like other

Dust storm in 1930s obscuring village. Saskatchewan Archives Board photo S-A295.

Saskatchewan housewives, learned that there was really no effective defence against them. Sheets hung over doorways, damp cloths along window sills, covers over exposed food, but nothing stopped the fine dust that permeated every aspect of their lives for years.

ENDNOTES

1. While Eggleston was in Limerick, Charles extracted a promise from him—that he would return and write a novel about the suffering in the South Country. Eggleston did, and *The High Plains* was published in 1938. But the story centered on southeastern Alberta where Eggleston's parents had homesteaded on former range land and suffered the earlier drought in the years following the First World War.
2. Victoria Trust and Savings Company, *Thirty-sixth Annual Report* (Lindsay, ON: Victoria Trust and Savings Company, 1932).
3. Ibid.
4. Peel, "R.M. 45," 390.
5. John H. Archer, *Saskatchewan: A History* (Saskatoon: Western Producer Prairie Books, 1980), 217.

CHAPTER 13

THE LIMERICK PLAN

I n 1931, the crops withered in the fields from the American border north to the forest fringe, and all of Saskatchewan, individually and collectively, lost the last shred of hope that they were experiencing only a temporary setback. All now knew that the province was in the midst of a calamity of biblical proportions. In many areas, what had been privation became starvation. It was a desperate time that called for desperate measures, and the Anderson government responded.

The distribution of relief to starving families in the south, now enduring their third consecutive crop failure, had exhausted the financial resources of dozens of tax-strapped municipalities. In August, the provincial government stepped forward with the creation of an arm's-length, independent agency, the Saskatchewan Relief Commission, under the chairmanship of Regina businessman Henry Black. The commission identified ninety-five municipalities that had suffered three successive crop failures, seventy-seven that had suffered two failures, and another sixty-eight experiencing their first failure.[1] Against all the evidence, both Anderson and Black harboured hope that the commission would be needed only over the winter of 1931–32. It was not to be.

Relief was granted in the form of a repayable loan on the theory that it would remove the stigma of charity and make it easier for need-

Settler family on the move near Jackfish Lake, 1933. Charles facing camera.
Saskatchewan Archives Board photo F544-11-50-1.

ful families to take advantage of the program. Not surprisingly only a minuscule amount of relief debt was ever repaid.

The governments in Ottawa and Regina recognized the tragic mistake of allowing non-selective homesteading of former rangelands in the southwest and began subsidizing settlers to vacate and try again in the northern regions. Assistance was granted to seventy-eight settlers in 1930, and in 1931 a total of 1,000 took advantage of the program.[2]

Beginning again along the forest fringe was not an idyllic solution, and many undercapitalized settlers failed again. The problems were eloquently described in a letter to Henry Black at the Relief Commission written in June 1932 by an acquaintance struggling near the new community of Pierceland, in the far northwest of Saskatchewan.

How I got up here in Township 61-26-3 is a chapter all by itself. It might sound something like the Children of Israel trekking out of Egypt but Moses is not at hand and the Red Sea refused to part its waves. In short, I am what is generally known as "homesteader" in these parts, as a two-headed billy-goat who had signed on the dotted line, promising to love, cherish and greatly improve 160 acres of fairly heavy timber, mosquitoes and sand-flies (with the aid of none save Divine Providence) but one who unfortunately forgot to leave his stomach and the attendant occasional desire for food in the so-called dried-out areas of Southern Saskatchewan. If you were to put yourself in my position you would readily understand what is worrying me—with nine of us all told, on land 55 miles from railway—no cow, no garden stuff until possibly this coming fall—nothing except four uncompleted buildings—just a strong desire to continue to live. This week the Relief Officer left me an order from your commission covering two months' period amounting in all to $11.35 or $5.66 a month. This figures out as considerably less than one cent per meal for each of us. When one has absolutely no other substance except this order (as is the exact truth in my case) it is practically impossible to live for that length of time, especially when groceries are purchased at prices not at all similar to city groceries. Strange as it may seem, I have lost exactly 54 lbs. in weight up here due largely to undernourishment and hard work on the dear old homestead and believe me, if this continues, I will be able to make Mr. Mahatma Gandhi look like the fat man in a circus. ...

We only had meat about three times all winter. For the past two weeks we subsisted chiefly on bread and water. I was clearing 10 acres of land at the time and had to quit, as I played out on this rigorous diet.[3]

When the miserable 1931 crop was harvested, some of the figures were astounding. The heavy soils of the Regina Plains and Weyburn in Crop District No. 2 produced a paltry average of 1.8 bushels per acre. District No. 3, running from the American border to the South Saskatchewan River and including Limerick, averaged 3.1 bushels per acre. The best-performing southern district was No. 4, in the extreme southwest, which managed 5.7 bushels per acre.[4] Throughout the province, many failed even to recover their seed.

In Ottawa, the government of Prime Minister R.B. Bennett, recognizing the hardship in the west, passed legislation on August 3, 1931, providing for the payment to farmers of an additional five cents per bushel of wheat produced.[5] Unfortunately, the subsidy was of minimal assistance in the hardest-hit areas, where almost no wheat qualified.

Charles, looking after 1,000 mortgages in 1931, had a very difficult year. Victoria Trust and Savings Company had a total Saskatchewan investment of $1.5 million in 585 mortgages scattered from Manor in the southeast to St. Walburg in the northwest, and from Shaunavon in the southwest to Nipawin in the northeast. Charles also supervised 420 Canada Life mortgages concentrated in a block south of Old Wives Lake and between Assiniboia and Kincaid, a total investment of another $1 million. Visiting those claims, Charles drove approximately 20,000 miles that summer, an arduous achievement even with a new car, considering the state of Saskatchewan roads at that time. Almost every mortgage that he reported on had fallen into arrears from a comfortable situation in 1929, when all of his Canada Life loans were current, as were all of Victoria Trust's southern mortgages.

In the early part of December 1931, the Board of Trade of the Village of Limerick called a meeting of the farmers of the district to see if anything could be accomplished by a frank discussion of the circumstances of the time. In the letter which called the meeting, one of the subjects suggested for discussion was "Are we in favour of repudiating our debts?" The response to the meeting took the Board of Trade somewhat off their feet, as not even standing room was available in the theatre. A Resolution Committee of seven was appointed, of whom five were farmers.

A resolution was introduced that deplored advocating the repudiation of debt and stated, in part, "We hereby declare to the world at large that we, for our part, intend to continue in the traditions of our race, and to honour all our obligations, whether individual, municipal or otherwise, to the full extent of those obligations if possible, and otherwise, to the fullest extent which future conditions in our country will render possible."[6]

It is not difficult to see Charles' hand in drafting the resolution, which, Charles reported, passed unanimously by a standing vote. Undoubtedly he was behind another resolution adopted expressing concern that 8 percent was "a higher rate than the [farming] industry can stand" and requesting that the lending and credit institutions conduct a survey "of the present conditions surrounding Western agriculture" so that farmers and lenders "may more fully understand each others' difficulties."

Then the meeting turned to consideration of a proposal that had originated with Harry Drope, of Regina, Andrew Hosie's partner in the firm Drope & Hosie, and then advanced by Charles and that came to be known as "The Limerick Plan."

Major (later Colonel) Andrew Hosie.
Photo in posession of the author.

Briefly put, the plan would obviate the unseemly scramble among creditors for the meagre returns of a harvest and instead provide for the orderly handling of farmers' affairs. A group of some 100 farmers would form a loose partnership under the suggested name Limerick Bureau of Farm Management and engage an administrator who, with limited secretarial help, would take charge of each farmer's total harvest receipts and distribute the funds appropriately. This would alleviate the farmer's stress and ensure fair and equitable debt retirement, with due consideration for the farmer's needs for his family's welfare and even repair, or replacement, of machinery. The cost of this administration was estimated not to exceed 1 percent of the expected farm income so handled. At least, it was suggested, "the scheme would cost less than the scores of collectors who will, in the absence of some such scheme, be busily engaged in our territory, and we believe, in the long run, the cost of these collectors goes back to the farmer."

The plan received the solid endorsement of that December farmers' meeting in Limerick. In the desperate winter of 1931–32, when the problems were vast and possible solutions scarce, it caught a good deal of public attention, much of which attached to Charles as its main proponent and spokesman. As he accepted invitations to speak to farmers' groups on the subject, he became well known across the province as someone who perhaps could make a useful contribution to alleviating the escalating level of farm debt.

As the calendar turned into 1932, there was no slowing of the never-ending drop in the prices of farm products. On February 23rd, in response to a request from Dr. Thomas Donnelly, MP, Charles identified that day's prices in Limerick: Number One Northern wheat, forty-one cents a bushel; one dozen eggs, twenty-seven cents; one pound of dairy butter, twenty-five cents; live cattle, nine cents a pound. As ugly as those prices were, they were headed still lower as returns continued in freefall throughout 1932.

The swelling mountain of farm debt was rumbling like a dangerous volcano and began to attract a good deal of public attention. On March 6, 1932, the Saskatchewan legislature appointed a committee of eight MLAS "for the purpose of consulting with the creditor and debtor classes of the Province, with the object of evolving some practical scheme for the re-arrangement or re-adjustment of indebtedness."[7]

The committee called a number of witnesses, including Charles on March 16th. He introduced himself to the committee as something of a hybrid, both a loan company man and a farmer representative, a concern that would arise a few years later in a larger sphere. He explained his work on behalf of two loan companies.

[It] gives me direct contact with slightly over 1,000 mortgages in the province, and the contact is a personal one as I do practically all the driving myself, my mileage during the summer of 1931 being around 20,000. I feel that this gives me an experience of the present situation in different parts of the province which is perhaps not available to everyone. It would be hard to say exactly in what capacity I appear before the committee. Notwithstanding my connection to two mortgage companies, I hold power of attorney from many farmers and am the personal agent of many more. If I make the claim to appear before this committee as voicing the opinions of the farmers of my own district in the south, I feel sure that I will not be challenged by these farmers.

Charles explained to the committee that, under the Limerick Plan, when the proceeds of the 1932 crop were realized, they would be turned over to the Bureau of Farm Management and distributed equitably among the farmers' creditors, retaining enough to pay harvest expenses and carry on in the next year.

I was on the stand for quite a period. After I had given what evidence I had to offer, I was subject to cross-examination for some time, particularly at the hands of one Jacob Benson of Last Mountain who had a fairly well-marked reputation as being of the extreme left. He had not entirely liked my evidence, and he asked me, "Mr. Wilson, do you not sometimes feel in danger on the road allowances, at the hands of farmers who are fed up?"

I replied to him, "Mr. Benson, I am conscious of being in danger on the road allowances occasionally, but not from the quarter you indicate. If I were to accept even one-half of the invitations which are given to me to go in the farm home and accept the hospitality which is offered to me, I do consider that my life might be in grave danger in a month or so. But apart from that, I have not the slightest feeling of uneasiness. I am met with courtesy and friendliness in every farm yard and find the farmer willing to sit down with me at his table and discuss this problem for which we both desire a solution, in soberness and good humour.

His reply to me was, "I think that must be blamed simply on your personality. Other people do not get the same impression."

The committee heard differing views from other witnesses. E.J. Davis, from Truax, who, like Charles, wore two hats, being both a farmer and a merchant, thought that the situation called for more desperate measures, including the forced liquidation of some farmers for whom there was no reasonable hope of making payments on their liabilities. Davis was reported as being of the "opinion that the morale of the people in his district was breaking, that there was open talk of revolution, and unless swift action was taken to bring about readjustment, serious developments could be looked for."[8]

George F. Edwards, of Markinch, a prominent farm leader and past president of the Saskatchewan Grain Growers Association, appeared for a second time before the committee, called back to respond to members' questions. He agreed that morale among farmers was low, "lower than he had ever known it, and little hope lay in the next years crop." Edwards was of the opinion that an adjustment of farm debt was needed to prevent collapse of the agricultural industry, and "a moratorium of debts was not unthinkable."[9]

Charles continued with his efforts to promote the Limerick Plan. Invited to speak to a gathering of farmers at Craik, he addressed the problem of his dual allegiance.

Perhaps it would be proper here to say a word to those of you who do not know me, touching on my qualifications to advance any views whatever on this difficult and complex question. First of all, I am the son of an Irish tenant farmer and the descendant of a long line of Irish tenant farmers. If any of you came to this country from the Irish farm, you will understand something of the devotion, amounting almost to passion, which the Irish people have for land and agriculture. I have my full share of that passion, and I consider agriculture to be not only the oldest but also the most honourable profession in the world in this or any other age.

The farmer is a producer of wealth and not simply a maker of money, and the wealth which he produces from his contact with the soil is perhaps more universally distributed than any other form of wealth. When the farmer is harvesting reasonable crops, he is giving employment to countless thousands, not only in his own community, but in communities far distant from him.

But the Limerick Plan was stillborn, doomed to failure from the start. It was founded on the expectation that 1932 would see something of a return to normal conditions, with a crop that would produce at least some revenue, enough to require careful distribution. In the spring of 1932, conditions throughout the wheat belt did promise some improvement, but the plan withered with the grain fields in the heat and drought of that summer and the despair that attended what has been determined to be the lowest wheat price in 300 years.[10]

That price was reached on December 16, 1932, when Number One Northern brought a paltry thirty-eight cents per bushel—at Fort William. The elevator, or street price, at most Saskatchewan communities of the more common grade of Number Two was roughly twenty cents.

A 100-bushel wagonload of Number Two Feed oats might bring a farmer only $2.50 in cash.[11]

Although crop yields in 1932 did improve somewhat over the disaster of 1931 (they could hardly have been worse), the ruinous prices left net farm income in the red. There was no revenue to distribute. There was nothing for taxes, nothing for loan companies, nothing for banks, nothing for merchants, many of whom were still carrying grocery accounts to keep suffering families in food, nothing for machinery upkeep or repair, nothing for the farmer, and nothing for his wife and children. Nothing, that is, but more relief.

The hope that the Saskatchewan Relief Commission would be a one-year program faded along with the Limerick Plan. Premier Anderson and Henry Black had no choice but to continue providing meagre succour to the distressed families of drought-ravaged rural Saskatchewan. The relief rolls did shrink a bit from the calamity of 1931, when 304,410 residents received assistance, just about one-third of the total Saskatchewan population of 921,785. In 1932–33, the second year of the program, 120,875 individuals were granted relief. Even after two years of operation, the commission was unable to close its doors, and in 1933–34, its third year, assistance was distributed to 214,742 destitute Saskatchewan residents.[12]

ENDNOTES

1. Gregory P. Marchildon and Don Black, "Henry Black, the Conservative Party, and the Politics of Relief," *Saskatchewan History* 58, 1 (2006): 4.
2. *Saskatchewan Journals of the Legislative Assembly*, Vol. 30, 1932, 275.
3. Marchildon and Black, "Henry Black," 17.
4. *Saskatchewan Journals of the Legislative Assembly*, Vol. 34, 1936, 13.
5. An Act Respecting Wheat, Statutes of Canada, 21–22 George V, c. 60.
6. The Limerick Plan, appendix A.
7. *Saskatchewan Journals of the Legislative Assembly*, 1932.
8. *Leader-Post*, March 16, 1932.
9. Ibid.
10. John H. Archer, *Saskatchewan, A History*, 1980, Western Producer Books, Saskatoon. 216
11. Fairbairn, *From Prairie Roots*, 113.
12. Marchildon and Black, "Henry Black," 4n2.

CHAPTER 14

THE FARMERS' CREDITORS
ARRANGEMENT ACT

In 1932, Charles squarely faced the bleak reality that he had not escaped the fiscal famine sweeping the west, and he knew with certainty that his financial security had evaporated along with that of so many others. One spring day walking down to his office, he stopped at the Bank of Montreal and picked up his safety deposit box containing most of his investments and securities. He asked Lizzie Allonby to shield him from visitors, closed his office door, and began a ruthless appraisal of his assets, seven quarter sections of land subject to mortgage, plus stocks, bonds, hail notes, and the like. It had all once amounted to a considerable sum, but when Charles completed his assessment he knew that he could no longer meet his liabilities. And there was one huge liability. The Bank of Montreal was holding his note for financing the hail insurance premiums that remained unpaid when the notes could not be collected after failed harvests.

After the first year of that failure, though Charles thought he should discontinue carrying notes for the hail premiums, he allowed his bank manager to convince him that he should stay with the business one more year. The result was an outstanding note of $35,000 at the bank (equivalent to more than $575,000 in 2012 dollars), and, on the other side of the ledger, the assets that Charles had just evaluated were close to worthless. He was seriously insolvent.

There would be no more trips to the Empress Hotel in Victoria (there had not been any since the spring of 1929). Although likely unknown to Charles, the Empress was also suffering as the Depression spared no victims. Guests became so scarce that they were outnumbered by employees, and many of those guests were permanent residents in difficulty as their retirement portfolios shrivelled. The Empress accommodated them by charging a mere dollar per day for its rooms on the top floor and turning a blind eye when formerly well-off dowagers smuggled in hot plates and groceries.[1]

Charles went home early the afternoon of the day that he accepted the loss of the financial security he had worked so hard and successfully to achieve, his mind awhirl with worry. He stood in the dining room and looked over at the house next door, the residence of his good friend, Limerick veterinarian Dr. Herb Gordon, and realized that he had never taken the time to visit "Doc." His daughter, Moira, raced in from school and, delighted to see her father, leaped into his arms, nearly knocking Charles down. She was thirteen years old, and as Charles set her back on her feet he was astonished to see how she had grown. He had been so engaged in work and politics that he had neglected his family and friends. And a fourth child had been born that spring, another son, Garrett.

Almost all that had been accomplished by the years of fierce devotion to his business, and to the Liberal Party, had disappeared like smoke, and Charles realized that he had allowed his life to fall seriously out of balance. That afternoon in 1932 he resolved to restore equilibrium to his personal affairs. Walking around his mostly unimproved yard, the thought of a garden entered his mind. As a boy, he had learned something of growing flowers from a neighbour, and County Wicklow was known as the Garden County. That thought progressed into action, and within a few years the Wilson yard had developed into a landmark garden.

But there was still the overwhelming problem of the ever-growing farm debt that was crushing the life out of the west, and Charles was inescapably in the middle of the maelstrom. As bleak as his own financial circumstances were, he knew many others, a great number of them personally, who were literally struggling for their survival and that of their families. Three of the seven quarter sections that

Charles owned, all mortgaged for amounts now well above any value they might hold, were near the tiny community of Stonehenge, about ten miles southeast of Limerick. They were rented by Ted and Fannie Oancia on the standard one-third crop rental arrangement. Both Ted and Fannie had come from Romania as children with their parents. They had made their home on the land rented from Charles, and in 1932 they had six young children to feed and clothe. Four more would arrive over the next ten years.

Charles and Ted, both immigrants from Europe and both with peasant backgrounds, had the utmost respect and affection for each other, particularly in the matter of integrity. All during the difficult years of poor crops and worse prices, they sorted out their affairs and kept the farm going without a cross word between them.

For his 1931 annual report to Victoria Trust and Savings Company, Charles included the resolution from the Limerick Board of Trade meeting in December, and Victoria Trust's president, William Flavelle, mentioned it in his presentation to shareholders, some of whom were concerned about the western investments.

Flavelle noted that the Limerick meeting was

held in the very heart of the dried out area and … was called to protest against the propaganda of certain elements of the population who were advocating a reduction of 25% on all debts by Legislative enactment. The meeting was very largely attended, and after full and free discussion, the following resolution was presented:

We hereby declare to the world at large that we for our part intend to continue to honour all our obligations, whether individual, municipal or otherwise, to the fullest extent.

The resolution was passed by unanimous standing vote and I believe this fairly represents the spirit of the great majority of Western Canadians.[2]

The loss of another crop in 1932, and the death of the Limerick Plan, began to work a change in Charles' opinion about how best to deal with the calamity.

Up to and including 1928, our farmers, like farmers elsewhere in the world, planted a crop in the spring and confidently expected to harvest it in the fall, and they were not disappointed. Occasionally hail or frost or rust might intervene, but the fact remained that the crop grew, and of course we expected that such a condition would always continue. Then, starting with the year 1929, we entered a period where for almost ten years we planted a crop, and that was almost the end of the matter. There was little harvest.

Something would have to be done to deal with the relentless growth of farm debt that was smothering the west, and that something might have to be as radical as the repudiation, or disallowance, of some of the unpaid interest, perhaps even a portion of the principal. It was essential to keep farmers on the land. To ensure this, creditors would have to make some concessions.

The report of Victoria Trust's 1932 annual meeting, held on February 7, 1933, addressed shareholders' concerns with the security of their investments in the company. T.H. Stinson, KC, vice president, told the meeting that

We are still in the days of stress and strain and depression. Our financial structure is being scanned, examined, weighed, praised and damned as never before and fortunately our financial institutions have stood up in the front line of all our activities. This is something to be proud of in this country. In four years of depression which has been more profound and extensive than ever before in the history of the world, not a bank has failed—not a Trust company has failed—not a Loan Company has failed. Our great neighbour to the South has had over five thousand bank failures together with a great number of other financial failures. ...

Mr. Wilson, our Western Manager, in his report at the end of the year said: "I look upon the expressed and keenly felt desire of our people to retain their homes and to work out the problem of debt in a spirit of fair play between themselves and their creditors as one of the most constructive and hopeful factors in our present outlook."[3]

George Edwards, who had appeared with Charles before the legislative committee, was firmly of the view that some reduction of farm debt was unavoidable if the agricultural industry was to be preserved. And Edwards was now a Liberal. The collapse of his Progressive Party had cast him loose, and he had chosen the Liberals as the party most closely aligned with his beliefs.

His decision to become a Liberal had not been quick or easy and was reached only after some serious thought and some unusual seduction. A prominent and highly regarded leader in the agrarian community, Edwards had been, among other offices, president of the Saskatchewan Grain Growers Association and president of the Canadian Council on Agriculture. James Gardiner, Saskatchewan Liberal leader and just-out-of-office premier, was anxious to recover the farm support that his party had lost to the Progressives. Fully aware of Edwards' stature, Gardiner approached Edwards to join and openly support the Liberal Party.

Edwards also knew that he had a strong following and told Gardiner that he would not agree to support the Liberals unless he first had an opportunity to discuss policy with Mackenzie King, also just out of office, having lost the June 28, 1930, election to R.B. Bennett's Conservatives.

In a remarkable acknowledgement of how tall Edwards stood in the western farm world, the Liberal Party purchased for him a train ticket to Ottawa. There he lunched with Mackenzie King at Laurier House, King's residence, and spent the afternoon in deep discussion of farm policy.[4]

The effort was successful as King convinced Edwards that the Liberal Party was closer to his views than any other. Edwards returned to Saskatchewan, wrote a number of articles for the daily and weekly newspapers explaining his choice of the Liberal Party, and openly campaigned for the Liberal candidate in a December 23, 1930, provincial by-election in Estevan.

The Saskatchewan Liberal Party needed a proposal for dealing with the farm debt crisis for its platform for the election expected in 1933 or 1934. It was natural that Gardiner would ask Edwards and Charles to sit down and work up such a plank. Edwards went down to Limerick, and Charles drove him through the South Country, where he knew so many of the struggling farmers and their mortgages. The two men hit it off and in the process of working together on the farm debt problem developed a close friendship that lasted until the end of their days.

Edwards was a proponent of what was somewhat cautiously called debt adjustment, actually a reduction, or writing off, of some portion of both interest and principal. On behalf of the Rural Municipality Association, he had first presented such a scheme to the provincial government during the agricultural difficulties following the First World War. By 1933, Edwards had little difficulty convincing Charles that there could be no other solution to the current crisis.

But 1933 was a year that started out full of promise. On May 27th, Charles reported this in his diary: *Plenty of moisture—in fact we are complaining and would like some fine weather for two weeks to finish seeding.*

On June 7th: *Heavy rain early this morning. A lot of water fell. Regina, Wood Mountain, Assiniboia all report good rain. Crop prospect excellent, except for hoppers.*

On June 23rd: *Went to Woodrow. Warm and dry. Crop west of Wood River pretty well gone.*

Seriously insolvent, and with a much reduced income that forced a tight budget, Charles still found room for an adventure that summer, perhaps related to son Kevin's graduation from College Mathieu at Gravelbourg. On July 5th, Charles and Kevin drove to Regina, where they joined Andrew Hosie and another friend for a trip to the Chicago World's Fair. The trip required eight days of driving for six days at the fair.

Hosie had been serving as chairman of the Saskatchewan Debt Adjustment Bureau since 1930 and was a strong Conservative. He and Charles did not allow their political differences to interfere with the close friendship that lasted their lives, and certainly many of the long hours of driving to Chicago and back must have been devoted to the farm debt problem in Saskatchewan and how best to deal with it. Subsequent events suggest that they influenced each other's philosophy.

Hugh McLaughlin and Wilson family. Photo in posession of the author.

As Charles noted in his diary, in early June there were prospects of a decent crop in 1933, but by the time he and Andrew returned from Chicago the crop was lost again, this time due as much to a grasshopper invasion as to drought. Grasshoppers had been an increasing problem for at least two years, and the egg census in the fall of 1932 had warned of a serious infestation to come the following year. Hundreds of poison bait mixing stations were manned, and the local hatch was considered to be under control when an airborne invasion of uncounted millions of the insects arrived from the United States and overwhelmed the defences in Saskatchewan and Alberta. Some $30 million worth of crops were lost, and the prognosis for 1934 was bleak.[5]

Charles' diary records almost every day of July 1933 as *hot and dry.* On July 25th, Charles set out with Hugh McLaughlin, vice president of Victoria Trust, on a tour of inspection of the company's Saskatchewan mortgages. It was a "See Saskatchewan" tour that covered the province from Manor in the southeast to Meadow Lake in the northwest. Starting from Regina, they drove to Findlater, Regina Beach, Balgonie, Cupar, Craven, Weyburn, Arcola, Kenosee, Manor, Lampman, Ogema, Limerick, Glentworth, Fir Mountain, Macworth, Coderre, Mortlach, Moose Jaw, Tuxford, Eyebrow, Central Butte, Lucky Lake, Strongfield,

Broderick, Ardath, Rosetown, Plenty, Wilkie, Denholm, North Battleford, Glaslyn, Meadow Lake, Saskatoon, Watrous, Wolseley, Fort Qu'Appelle, and back to Limerick.

On his return home on August 10th, Charles recorded this: *Hot and dry every day. Hoppers all over us.* Four days later, on August 14th: *Weather has been hot and dry without a change since early July. Crops are shot over a huge area.*

The next day Liberal leader Mackenzie King, on a tour across southern Alberta, Saskatchewan, and Manitoba, paid a visit to Assiniboia, travelling from Moose Jaw in the business car of the CPR superintendent. His diary reveals his impression on arrival at 1:45 p.m.:

At Assiniboia was given a very fine reception at the station— hundreds of people old & young. The young people all looked so fresh & clean & fair. I have never seen a nicer looking lot. I shook hands with most of them, then went to Dr. Gemmel's residence, preceded by a band & underneath the streamer of welcome. These people have suffered terribly, this is the 5th year of drought—yet they were cheerful. Certainly nothing could have been more cheerful than the welcome this morning, all so prettily dressed & smart & clean. I pray God I may be able to bring a real message to them from Him.

The political rally in the evening was much longer than would be acceptable today:

The meeting was on at 8, a very large crowd in a very warm rink. Spoke for about 1 ¾ hours. Dr. Donnelly took up a good deal of time at the beginning & was pretty extreme in what he said. I reviewed the Cons. position re tariff—the Liberal program & dealt at some length with the C.C.F. Was given a splendid reception. Must say the heroism of the people in these parts appeals deeply to me, their courage & cheer in face of 5 years of drought is beyond words.[6]

After the meeting, King returned to the Gemmel residence to confer with a delegation of local Liberals, including Charles, who advanced the claim that their MP, Dr. John Donnelly, be considered for inclusion in the federal cabinet when the Liberals returned to power. Charles took the opportunity to press the urgency of dealing with the farm debt crisis, threatening to empty much of the region.

After that depressing summer of 1933, Charles was ready to agree with George Edwards that some form of farm debt reduction was the only answer to the crisis facing rural Saskatchewan. The results of his collaboration with Edwards, benignly described as debt adjustment, were approved as a supplementary platform resolution by the Council of the Saskatchewan Liberal Party in January 1934 and added to the party's election platform. It called for the creation of as many "debt adjustment tribunals" as needed "to be available to all resident debtors. The services of such Tribunals shall be free."[7]

The tribunals would "make a full and complete investigation of the affairs of such debtors by ascertaining the value of all assets such as land and equipment, the total of all liabilities and the manner in which they were incurred." Then, "after due consideration" of all factors, the tribunals could order "a fair and equitable adjustment or reduction of such debts, both as to principal and interest."[8]

There it was. Debt reduction. That meant writing down mortgage and other loans to a level that could be supported and repaid with the much lower farm revenues that were the new reality in Saskatchewan in the 1930s. It was a bold initiative for the day.

Also in the Liberal platform was a proposal to limit the personal covenant in mortgages, the individual commitment to repay the original loan that survived even foreclosure if sale of the seized lands did not produce enough to retire the liability. That issue had been debated in the legislature during the 1933 and 1934 sessions, but the Anderson government had declined to adopt the provision.

Even if the Liberals won the election expected to be called in June, which looked more than probable, there was the looming question of how much constitutional authority the province had to enact effective legislation in the field of debt reduction. The subject smacked of bankruptcy and insolvency, areas specifically reserved to Parliament by the British North America Act of 1867, then the Constitution of Canada. The area of property and civil rights was assigned to the provinces but did not appear to provide all the jurisdiction that the problem required.

Andrew Hosie at the Debt Adjustment Board recognized this inadequacy in his annual report of December 21, 1933:

> Due to the hopeless financial position of many farmers we are of the opinion that the Dominion Government should be asked for legislation which would permit farmers to make assignments under bankruptcy without being subject to the present Dominion Bankruptcy laws which are not suitable to farmers. The present procedure is a tedious, lengthy and expensive one.
>
> During the last two years only twelve farmer bankruptcy cases have been handled by the board, which we believe has handled practically all cases. We are of the opinion that individuals and the province would benefit if arrangements could be made for more efficient farmer bankrupt legislation. Bankruptcy, failing a writing down of debt, or a substantial improvement in values of agricultural products, is the only solution for many of our farmers.
>
> Every effort has been made to settle differences of opinion between debtor and creditor on an amicable basis and to effect adjustments between debtor and creditor, but our efforts have been handicapped due to the fact that most creditors feel that the proper time to adjust debts is when the debtor has some cash to bargain with. Creditors as a whole fail to recognize that individuals cannot achieve the same results unless they have some definite hope of bettering their position.
>
> Many debtors feel that even if there is a temporary improvement in conditions their creditor will be reluctant to make adjustments that will give them a chance to liquidate their debts within a reasonable length of time. Debtors feel that they should be enabled to face the future with a fair prospect of success. We regret to say that the action of certain creditors in cases where funds have become available, confirm the debtors' opinion.[9]

Hosie thus identified a constitutional problem that plagued the western farm debt crisis for several years. Any effective solution to the mountain of debt threatening to overwhelm the west would necessarily cross federal and provincial jurisdictions.

It is unlikely that his opinion reached the Prime Minister's Office in Ottawa, but on June 4, 1934, Prime Minister Bennett rose in the House of Commons and responded to the concern with the introduction of "a measure to facilitate compromises and arrangements between farmers and their creditors."[10] Not quite a month later, on July 3rd, the Farmers' Creditors Arrangement Act, 1934, received royal assent and came into being.

Bennett did hope that his new measures, including the Prairie Farm Rehabilitation Act, enacted on April 17th, would assist not only his electoral prospects but also those of his Conservative colleagues in Saskatchewan. There, on May 25th, Premier J.T.M. Anderson called the long-awaited election for June 19th.

ENDNOTES

1. Godfrey Holloway, *The Empress of Victoria* (Victoria: Pacifica Productions, 1968), 57.
2. Victoria Trust and Savings Company, *Thirty-Sixth Annual Report* (Lindsay, ON: Victoria Trust and Savings Company, 1932), 9.
3. Victoria Trust and Savings Company, *Thirty-Seventh Annual Report* (Lindsay, ON: Victoria Trust and Savings Company, 1933), 10, 11.
4. Edwards, "Memoirs of George F. Edwards," 27.
5. James H. Gray, *Men Against the Desert* (Saskatoon: Western Producer Prairie Books, 1967), 41.
6. LAC, Mackenzie King Papers, Diaries, MG 26, J13, vol. 63, 319–20.
7. Saskatchewan Liberal Party Supplementary Platform Resolutions, adopted by the Council of the Party, January 9–11, 1934, in possession of the author.
8. Ibid.
9. Andrew Hosie, "Report of the Commissioner of the Debt Adjustment Bureau," December 21, 1933, *Journals of the Saskatchewan Legislature*, Saskatchewan Legislative Journals, 1933.
10. House of Commons, *Debates*, June 4, 1934, 3637.

CHAPTER 15

DEBT ADJUSTMENT

Prime Minister Bennett's legislation was too little too late to be of any help to the Conservative/Cooperative government in Saskatchewan. In fact, the prime minister himself had become an anchor around the neck of the drowning Anderson government, and the name Bennett had evolved into an epithet. A commonly seen contraption in western Canada was an automobile converted into a horse-drawn conveyance by a farmer who could no longer afford gasoline. With the engine removed, the front wheels moved back, a wagon tongue affixed, and the unit pulled by two horses (usually), it became the ubiquitous and cynically named Bennett Buggy. In Saskatchewan, a more extensively renovated version had the entire front half of the car's chassis removed, leaving a two-wheeler known as the Anderson Cart. Wheat roasted and then brewed produced something called Bennett Coffee.

The voters of Saskatchewan were suffering and angry. In the election of June 19, 1934, they defeated the entire Anderson cabinet and every MLA on the government side. Fifty Liberals were elected, and James Gardiner returned to the premier's office. The remaining five seats in the fifty-five-seat legislature went to the Farmer–Labour group, the precursor of the Cooperative Commonwealth Federation (CCF). It would be nineteen years before another Conservative was elected to the Saskatchewan legislature.

Bennett Buggy driven by Mackenzie King, Sturgeon Valley, 1934.
LAC C-000623

Charles spent election day at Flintoft; it had now become a tradition
with him. This time there was no anxiety as Liberal Charles Johnson
easily won the Willow Bunch constituency with 2,448 votes over the
Conservative with 1,445 and the Farmer–Labour candidate with 1,219.

For Charles, there was a downside to the Liberal landslide in the
midst of the Depression. Widely known as a party wheel-horse, he
found himself so beset by job and favour seekers that, to secure some
peace and sanctuary at home, he had the telephone removed.

In Ottawa, the prime minister pressed forward with his new initiatives. The Farmers' Creditors Arrangement Act (FCAA) was quickly
proclaimed in Alberta, Saskatchewan, and Manitoba and later extended to all provinces. On July 24th, little more than a month after
his defeat as attorney general in the Anderson government, Murdoch
A. MacPherson, a highly regarded Regina lawyer, was recruited to
Ottawa to "assist in the organization and administration" of the
FCAA.[1] Bennett and MacPherson moved swiftly. By the end of 1934,
machinery was in place and operating in 130 judicial districts covering
all of Canada, and each province had its supervisory Board of Review,
consisting of a federally appointed judge of the Superior Court as
chief commissioner and one commissioner representing creditors and
another representing debtors.

Anderson Cart near the CPR station in Limerick, 1933.
Photo in possession of author.

The preamble to the FCAA read as follows:

> Whereas in view of the depressed state of agriculture the present indebtedness of many farmers is beyond their capacity to pay: and whereas it is essential in the interest of the Dominion to retain the farmers on the land as efficient producers and for such purpose it is necessary to provide means whereby compromises or rearrangements may be effected of debts of farmers who are unable to pay.

The procedure provided by the FCAA was the appointment in each judicial district (there were then twenty-one in Saskatchewan) of an official receiver to accept proposals for rearrangement of a farmer's debt. Once a farmer (or creditor) had filed a proposal, a stay of any further proceedings against him, such as foreclosure, went into effect. The official receiver would then meet with all creditors to determine if the proposal was acceptable. If it was not, then the file would move on to the Board of Review. The board would then work out an arrangement based on the capability of the farmer and the productive value of his farm.

Speaking to the Toronto Canadian Club on November 5, 1934, MacPherson explained that the FCAA recognized "two types of farmers in difficulty—the man who is so hopelessly involved that only through the Bankruptcy Court could he get relief and a fresh start, and the other man technically insolvent, with sufficient assets but in default and in the position where rearrangement of his debt was necessary."[2] In thus describing the purpose of the FCAA, MacPherson omitted any reference to debt reduction, whether of interest or of principal, still a tender subject in the corporate boardrooms of Ontario but an element that became critical in the operations of the Boards of Review.

Staffing the machinery of the FCAA in all nine provinces was a patronage bonanza for Prime Minister Bennett and the Conservative Party in mid-Depression. Commissioners for the Boards of Review, official receivers in 130 judicial districts, and supporting clerical and secretarial staff totalled hundreds of positions for loyal party supporters. MacPherson was up to the challenge.

When Prime Minister Bennett introduced the FCAA in the House of Commons on June 4, 1934, he quoted the figure of $726,026,500 as the total Canadian farm mortgage debt. That number was based on the 1931 census and, as he admitted, was seriously out of date but nonetheless illustrative of the magnitude of the problem (see Table 15.1).[3]

ESTIMATED TOTAL FARM MORTGAGE DEBT OF CANADA, BY PROVINCES.

PROVINCE	INDEBTEDNESS ON OWNED FARMS (Census Figures)	ESTIMATED INDEBTEDNESS ON RENTED FARMS	TOTAL FARM MORTGAGE INDEBTEDNESS
	$	$	$
Prince Edward Island	4,866,700	59,700	4,926,400
Nova Scotia	6,570,000	78,100	6,648,100
New Brunswick	6,485,400	99,300	6,584,700
Quebec	96,409,400	1,712,100	98,121,500
Ontario	199,755,100	10,560,800	210,315,900
Manitoba	59,223,400	7,934,100	67,157,500
Saskatchewan	175,770,300,	20,788,000	196,558,300
Alberta	107,519,000	12,542,100	120,061,100
British Columbia	15,177,200	475,800	15,653,000
Canada	671,776,500	54,250,000	762,026,500

Table 15.1 : Debates, June 4, 1934, 3649.

By the end of 1934, the situations of some 3,000 farmers across Canada had been considered under the FCAA by official receivers, and many satisfactory settlements achieved, but no case had yet reached a Board of Review.[4]

In Regina on July 19, 1934, T.C. (Tommy) Davis (see p. 227), the MLA for Prince Albert, was again sworn in as attorney general, the portfolio that he had held in the Liberal government that had fallen in September 1929. Davis was intelligent, personable, conscientious, and industrious, attributes that would carry him to high offices nationally and internationally.[5] When he again took up his old responsibilities that summer, he focused his attention on the debt problem that was threatening to engulf Saskatchewan.

Davis set about drafting a new Debt Adjustment Act for presentation to the first session of the new legislature, planned for the fall of 1934. He intended the new act to have teeth, a clear power of moratorium that would enable the board to cancel or reduce debt, with or without creditor consent. As the attorney general would later explain when describing the act to the legislature, "this law will give the Government and the Debt Adjustment Board complete control over the debt situation in this province."[6]

But first he had to find the answer to a thorny question: what gave his or any government the right to interfere with hundreds of thousands of contractual obligations entered into between willing lenders and borrowers all acting in good faith? Was the time-honoured principle of sanctity of contract to be abandoned?

When Davis stood before the legislature on November 26th, he presented his carefully worked-out rationale.

In the first place let me say that I am of the opinion that this Legislature is practically unanimous upon the necessity of debt adjustment, and that it is absolutely essential that legislation to provide a means of debt adjustment should be enacted at the earliest possible moment.

All are only too well aware that, at this time, we are passing through the greatest period of depression the world has ever known, a depression which of necessity, has been accompanied by a state of unusually depressed prices, particularly of agricultural commodities. Agriculture is the basic industry of Saskatchewan and, by peculiar coincidence, this period of depressed prices for agricultural products has been accompanied also by the severest and most intense drought that Western Canada has ever experienced. Then, too, associated with the drought has been the affliction of grasshoppers and other pests disastrous to agriculture. As a result the value of agricultural production, in many instances during the last few years, has been entirely wiped out or seriously curtailed.

The Debt Problem

DURING THE TIME ELAPSED SINCE THE DEPRESSION CAME upon us, and particularly during the period of drought, very little additional indebtedness has been incurred in this province for two very good reasons; first, everyone has been too busily engaged in trying to handle the debt heretofore incurred to give much thought to the taking on of additional burdens and, second (and a very excellent reason), as no one was prepared to lend anything, no one could go into debt even if he so desired.

The problem, therefore, with which we have to deal is largely a problem of debt incurred prior to the depression and prior to the drought period.

During this whole time, by reasons of depressed prices and lack of production through drought, there has been very little reduction in debt by payment, but, on the contrary, there has been a large increase through the addition of interest which our people, through no fault of their own, have been unable to pay.

These conditions have brought about decreased values of land, because, of necessity, the value of land is based on the value of the production thereof, and decreased production either in volume or amount brings about decreased land values. We have, therefore, had an ever increasing lowering of values of farm land. Hand in hand with this, there has been an ever increasing amount of debt charged against the land by reason of the increase of secured charges through addition of interest.

Farmer's Equity Shrinking

THE FARMER'S INTEREST OR EQUITY IN THE LAND IS THE difference between the value of the land and the amount he owes thereon; and, as the value of the land has become progressively less and the amount of the charges against it progressively greater, his equity, therefore, has become progressively less with the very grave danger of the farmer's equity being entirely eliminated by reason of the ever increasing approach of the value to the amount of the debt.

We must remember that these debts were incurred at a time of high price levels, and under conditions which those who borrowed the money and those who loaned it reasonably believed would continue. They believed, and reasonably, that the conditions which prevailed when the money was borrowed and loaned, would continue on an even keel during the period over which repayment of the debt would be made. In other words, no one anticipated the drastic drop in prices of farm commodities and the total lack of production in the drought areas of the province.

By virtue of these conditions, the question, therefore, arises as to who should bear the loss occasioned by these conditions—conditions over which neither the borrower nor the lender has had any control, conditions not due to fault on the part of either.

Under ordinary business conditions, the loss occasioned by inefficient management or other matters over which the borrower has control must, of necessity, be borne by the borrower; but when abnormal conditions are upon us, then the ordinary business rule must, in my opinion, be varied in operation to meet the exigencies and circumstances of the case.

Therefore, one of the major problems for settlement in this province is the question as to who should absorb the consequent loss.

It has been argued in this province that, when money was loaned on the security of land, the value of the land was ascertained and the lender entered into partnership with the borrower, and the respective interest of each in the partnership property (namely, the land) was in the same proportion as the amount of the mortgage bore to the value of the land.

There is not now and never has been any logic to this argument and I am certain that, if conditions were as good as they were some years ago, with prices and production as high as then prevailed, no person with a reasonable charge against his land would have been prepared to share with the lender of the money the proceeds of his farm operations on a partnership basis. This suggestion is only made after it becomes apparent that it might ease the burden if the relationship were treated on this basis.

The fact remains that, when money is borrowed on the security of land, the borrower, subject to payment of taxes and prior charges, gives the lender a first charge on the land for the amount of the mortgage plus interest at the agreed rate. Notwithstanding this relationship and notwithstanding the legal position which might entitle the lender to repayment of the whole of the amount advanced plus all interest thereon, yet, when we are faced with the economic conditions of the depression through which we have passed and are passing, and when we are faced with a national calamity such as the drought conditions which have prevailed in the province, we realize that this legal position must be altered.

Keep People on Lands

I THINK WE ALL REALIZE TOO THAT OUR LANDS HAVE value only so long as there is someone living on them, cultivating them, and producing something from them. Vacant land has no value except the possibility of the unearned increment which may come about as a result of the effort and production of others.

It is essential, therefore, in the interests of our people and of those who are interested financially in our province, that a condition be created which will permit our people to remain on their lands, pay their debts, and be once more happy and contented.

I am positive that, in the majority of cases, no one realizes better than the creditor class the truth of the statements which

I have made, with the result that every day there is going on in this province a steady stream of debt adjustment.

It has been said (and for some time I thought with possibly sound logic) that nobody could adjust debt or could reasonably ask for debt adjustment until he had something with which to adjust debts, that, until he had some money to pay to his creditors and at the time of payment ask for some adjustment. Further thought on this subject has led me to the conclusion that this position is not sound. We must, of necessity, maintain the morale of our people, and appreciate the fact that a man, who sees his debts ever increasing through no fault of his own and sees gradually slipping away from him everything he has in the world, has not the heart nor the spirit to continue the battle against the conditions which presently prevail.

If he is assured that, when the time does come, he can secure an adjustment and that an adjustment will be made, then he will tackle his problems with renewed heart and with renewed vigor.

Debtor Entitled to Consideration

IT IS, THEREFORE, ESSENTIAL THAT THERE SHOULD BE indicated to our people the fact that they are entitled to, and will receive, reasonable treatment in connection with the sharing of the loss, as between creditor and debtor, which has been occasioned through circumstances over which neither of them has had any control.

There are many cases where the amount of the secured debt, with addition of all interest, would not yet equal the reasonable value of the land and, under ordinary circumstances, the creditor might reasonably expect to receive the last cent, but in so doing he is imposing on the debtor the obligation of assuming the entire loss which has been occasioned by prevailing conditions which, as I have said, have been beyond the control of either party. This class is entitled to consideration by way of a sharing of the burden of the interest which has accumulated, and which has not been paid because of adverse circumstances.

Then there is the matter for consideration of the relationship between the secured and the unsecured creditor. It is natural, under ordinary conditions, that, as the debt to the secured creditor increased by the addition of interest, and the value of the land decreased, the chances of the ordinary unsecured creditor getting anything are correspondingly decreased.

I think, therefore, it is only reasonable that, in any system of debt adjustment, the entire loss should not be thrown upon the unsecured creditor; and while it is only reasonable to expect that he should bear a greater burden of the loss than the secured creditor, yet the latter should make some concession to the extent of helping the unsecured creditor in some degree.

Mutual Agreement Best

I THINK WE ALL AGREE, TOO, THAT THE BEST TYPE OF debt adjustment is one mutually agreed upon by the parties immediately concerned and not one brought about by compulsion. If compulsion must be used then it is only used because the parties cannot agree, and if they cannot agree then neither party is very likely to be satisfied with the decision of the body which holds the power of compulsion. All that I have said with respect to the rural residents of the province, of course, applies with equal force to urban residents.[7]

BUT THE GOOD INTENTIONS OF DAVIS RAN UP AGAINST THE new FCAA just being put into operation from Ottawa. Aside from the constitutional conflict between the provincial jurisdiction over property and civil rights and the federal jurisdiction over bankruptcy and insolvency, there were more practical problems. The FCAA, by title as well as mandate, was restricted to dealing with the debt of a farmer, defined as "a person whose principal occupation consists in farming or the tillage of the soil."[8]

Although Davis was prepared to leave the field of debt adjustment to the federal government, he was not willing to exclude from the process those thousands of Saskatchewan residents who were not farmers but who were equally staggering under the load of debt. The impracticality, and duplicated costs, of having a federal scheme dealing with farmers and a provincial scheme dealing with non-farmers, operating side by side, was obvious. On August 15, 1934, the Saskatchewan attorney general wrote to his predecessor in that office, M.A. MacPherson, then in charge of setting up the machinery of the FCAA.

Davis pointed out that, constitutionally, the federal and provincial governments could not both occupy the field of debt adjustment; he was prepared to yield the field if Ottawa would widen the FCAA mandate to cover all residents, but if Ottawa refused to do so "we would have to reserve to ourselves the right to object to the federal legislation and to step into the field and provide such methods as we consider best to cope with the situation."[9]

In his response, MacPherson took the position that by enacting the FCAA the federal government had already occupied the field of debt adjustment and that the province was thereby prevented from similar action. That was not good enough for Davis. On November 13th, he again wrote to MacPherson, advising him that he was just completing a new Debt Adjustment Act that would not go beyond the legislative powers of Saskatchewan but would not ignore the needs of the many non-farmer residents.

> Let me conclude by saying that I agree that the subject matter of this legislation is of too much importance to result in any quibbling between the two Governments, as to the doing of the job. We do not want to put the people of this province in the hypothetical case of a patient being seriously ill and needing an operation, and finding two opposing surgeons disputing as to which should perform the actual operation. Thus, while the dispute is proceeding, the poor, unfortunate patient dies.[10]

With that, the Saskatchewan legislature enacted the Debt Adjustment Act, and the two jurisdictions, federal and provincial, embarked on parallel courses. Provincial legislation was also enacted in Alberta and, to a lesser extent, Manitoba. Constitutional challenges and litigation followed.

In 1937, the Judicial Committee of the Privy Council, in London, England, very much the court of last resort, upheld the constitutionality of the FCAA. Provincial legislation in the field did not fare as well. In 1943, the Judicial Committee held that the Alberta Debt Adjustment Act, 1937, was *ultra vires*, beyond the competence of the provincial legislature, a decision that also put the Saskatchewan board out of business.

Until the courts ruled against their legislation, the provinces carried on with their Debt Adjustment Acts. Saskatchewan recruited George Edwards, the long-time proponent of farm debt reduction, to serve as a member of the Debt Adjustment Board, along with Edward Oliver, who had laboured as a one-man board all through the 1920s, and Newton C. Byers as chairman. Andrew Hosie, the former chairman, was a casualty of the political sea change that had taken place in Saskatchewan.

Byers and Edwards were called on to serve with another undertaking, the Debt Survey, 1934–35, an attempt to quantify just how much debt had been accumulated by the various public bodies in the province, such as municipalities, both rural and urban, and schools. Byers and Edwards served on a committee chaired by Minister of Education J.W. (Bill) Estey, a Saskatoon MLA just elected in 1934, another of the extremely capable members of the Gardiner cabinet who was destined for high national office.[11]

All the attention now being paid to the debt problem was not misplaced. The 1934 crop was another bust. The statistics were stark, giving the agricultural economists something to really sink their teeth into. One of them, Professor George E. Britnell, at the University of Saskatchewan, became nationally prominent with his studies of the Canadian wheat economy. From an article that Britnell published in 1936, we learn that the 1934 wheat yield across Saskatchewan averaged just 8.6 bushels per acre, slightly below even that of 1933 and the lowest since 1920. The cash value of that crop was barely one-quarter of the 1928 crop.[12] The four crop districts across the south of the province did not achieve a five-bushel yield, and the cash returns per acre were below the level at which a farm family might subsist without assistance (see Table 15.2).

CASH RETURNS PER ACRE OF WHEAT IN CROP-REPORTING DISTRICTS OF SASKATCHEWAN, 1930–34[*]

DISTRICTS	1930	1931	1932	1933	1934
South-eastern	$5.83	*$1.29*	*$3.63*	*$3.29*	$2.01
Regina-Weyburn	$4.56	*$0.11*	*$3.32*	*$5.13*	$2.01
South-central	*$3.25*	*$0.61*	*$2.37*	*$1.18*	*$1.28*
South-western	$5.74	*$1.60*	*$4.06*	*$1.32*	*$1.60*
Central	*$4.28*	*$2.66*	*$3.53*	*$1.93*	*$3.98*
West-central	$8.65	*$4.52*	*$5.39*	*$1.23*	*$4.33*
East-central	$6.82	*$3.57*	*$5.49*	$10.30	$10.12
North-eastern	$10.62	$7.79	$7.11	$7.10	$9.27
North-western	$13.07	$7.28	$6.65	$5.07	$10.12

Table 15.2: *Based on crop yields and farm prices as supplied by the secretary of statistics, Department of Agriculture, Regina.

Note: Italics indicate districts in which government assistance was necessary.
Sourc: Britnell, "Saskatchewan, 1930–1935," 150.

In July 1934, Charles again guided officials of Victoria Trust on a tour of the company's mortgages in Saskatchewan. On the 23rd, he met C.E. Weeks, managing director, and Wesley Walden, director, who had made it as far west as Arcola on their own. The day that they met was, Charles described in his diary, *extremely hot and dry*, a condition that lasted throughout the tour.

From Arcola, the three mortgage men visited Victoria Trust's loans at Whitewood, Wolsley, Balcarres, Lemberg, McLean, Regina, Cupar, Craven, Bethune, Regina Beach, Ogema, Limerick, Meyronne, McCord, Wood Mountain, Craik, Aylesbury, Elbow, Strongfield, Hawarden, Outlook, Rosetown, Plenty, Wilkie, North Battleford, Bresaylor, Edam, Meota, Lilac, Perdue, Saskatoon, Watrous, Tuxford, and back to Regina. The trip was a depressing and dismal exposure to mortgages in arrears, and why they were so, and why they were likely to stay that way, for the Ontarians to take back to head office in Lindsay.

Drifting soil near Lakenheath.
Saskatchewan Archives Board photo R-A4822.

Soil erosion in the 1930s.
Saskatchewan Archives Board photo R-A4665.

Soil erosion near Kisbey, 1936.
Saskatchewan Archives Board photo R-B9060.

The drought, crop devastation, and human misery in southern Saskatchewan once again received in-depth press coverage. In September 1934, D.B. McRae, editor of the Regina *Leader-Post*, and R.M. Scott, assistant agricultural editor of the *Winnipeg Free Press*, teamed up on a 2,100-mile tour of inspection of the drought regions from almost the Alberta border into Manitoba. Their reports, filed from fourteen communities, received wide attention and were, together with a summary, later published in a booklet, *In the South Country.*

On September 19th, McRae and Scott visited Limerick and district, where one of their interviews was with Charles, whom McRae knew as a fellow member of Regina's Assiniboia Club. They found a grim determination to carry on in spite of another almost complete crop failure, the fifth for the region. Less than one bushel per acre was the yield estimate for the Rural Municipality of Stonehenge surrounding Limerick, most of the loss attributed to grasshoppers and soil drifting. The municipality expected that 90 percent of its 450 families would need relief to get them through the winter. Yet humour prevailed. "Came into this country with nothing and still have it" was a common joke.

Premier Gardiner's Liberal government had terminated Henry Black's Relief Commission and turned frontline administration back to municipalities. McRae and Scott found that, under the new system,

the municipal secretaries are busy taking applications for relief. These will be scrutinized by the reeves and councillors, many of whom are on relief themselves. But they say they will get by with the bare necessities. Men come in and give statements of their crop returns, the number of horses and cattle, their gardens, and what they will need. It is pitiful to hear a man affirming that he had 125 acres of wheat acreage and harvested 60 bushels of wheat.[13]

In their summary, McRae and Scott addressed the thorny issue of debt. They had found general agreement that the debts piled up on account of the extraordinary public borrowing to make it possible to carry on during the last four or five years would have to be adjusted in some way to bring them into line with prospective income even if good crops and fair prices were again realized. They said that the sooner the adjustment was indicated the better it would be for the morale of the people and the greater the prospect of the people rallying with new courage to carry the adjusted "load."[14]

When *In the South Country* was published, Premier James Gardiner provided a foreword in which he outlined the uniqueness and gravity of the problem of the drought area of Saskatchewan:

Wheat has brought more new money into Canada during the past 30 years than any other commodity. Saskatchewan has produced more of that wheat than the other eight provinces combined, while south-western Saskatchewan, now suffering from severe drouth, alone has produced twice as much wheat over any 20-year period as all the rest of the province. If kept there, the inhabitants will produce the same results in the next 20 years. The national question is not how to get them out of there but how to keep them there, if not on the lands now occupied, on better lands in the same part of the province.

In 1922 that portion of Saskatchewan, which today is drouthstricken, produced 150,000,000 bushels of wheat, while in 1928 the yield was 165,000,000 bushels, which is only 35,000,000 bushels below Canada's entire export quota, fixed by agreement among the wheat producing countries at last year's world economic conference.

In the five years between 1922 and 1928, south-western Saskatchewan produced enough new wealth to wipe off the entire indebtedness of every kind in the whole province.[15]

In the federal government in Ottawa, in the provincial government in Regina, and in the press, regionally and nationally, the concept of debt reduction, disguised as debt adjustment or not, had gained acceptance. And the machinery had been put in place to bring it about. What remained was how to make it happen to an extent that would have real beneficial effects on the farms and in the villages and towns of southern Saskatchewan.

During the winter of 1934–35, those farms, villages, and towns faced even more distress and misery than during the previous bleak winters. The Saskatchewan Voluntary Rural Relief Committee distributed its largest amount of assistance to date: 329 freight car loads of vegetables and fruits, 100 carloads of coal for schools, and more than sixty tons of clothing.[16]

Charles' last diary note for 1934 was a bit wistful: *Closing the year very tired physically and mentally. It has been a tough one. We are optimistic for 1935. A nice amount of snow on the ground and some moisture in the fallow. Plenty of rough feed and generally the farmers in better shape than for many years. We need a crop. Everyone is fairly well worn out with anxiety and overwork. Whatever criticism I may have incurred, I feel I have done a good and constructive year's work.*

ENDNOTES

1. On April 1, 1935, MacPherson's duties were changed to "assist the Board of Review in each province." His salary remained at $500 per month. LAC RG 19, Finance, Vol. 427. file 105-1A-12.

2. M.A. MacPherson, cited in J.F. Booth, "Measures for the Relief and Rehabilitation of Agriculture in Canada," *Journal of Farm Economics* 17, 1 (1935): 82.

3. *Debates*, June 4, 1934, 3649.

4. Booth, "Measures," 107.

5. Davis went on to become a justice of the Saskatchewan Court of Appeal, then deputy minister of war services in Ottawa, high commissioner to Australia, and ambassador to China, West Germany, and Japan. He retired in 1957 to Victoria, where he died in 1960.

6. *Journals of the Legislative Assembly*, November 26, 1934, 23.

7. Saskatchewan Legislative Journals, 1934, 5.

8. The Farmers' Creditors Arrangement Act, 1934, ch. 53.

9. Saskatchewan Legislative Journals, 1934, 18.

10. Ibid., 22.

11. In 1939, Estey succeeded Davis as attorney general, and in 1944 he was appointed a justice of the Supreme Court of Canada.

12. G.E. Britnell, "Saskatchewan, 1930–1935," *Canadian Journal of Economics and Political Science* 2, 2 (1936): 143.

13. D.B. McRae and R.M. Scott, *In the South Country*. (Saskatoon: *Saskatoon StarPhoenix*, 1934), 16.

14. Ibid., 40.

15. In ibid., 4.

16. Britnell, "Saskatchewan, 1930–1935," 156.

CHAPTER 16

1935

In early January 1935, Prime Minister R.B. Bennett startled Canadians when, in a series of five half-hour radio broadcasts, he explained his conversion to un-Tory-like government intrusion into the nation's economic affairs. The Depression had brought about such conditions that reform was needed, he said, "And, in my mind, reform means Government intervention. It means Government control and regulation. It means the end of *laissez faire.*"[1]

The "reform" that the prime minister had in mind was a series of legislative enactments to follow, such as creation of the Bank of Canada, that had profound effects on Canada's social order for years to come. Some legislation, such as the Farmers' Creditors Arrangement Act, had been passed the year before. The Prairie Farm Rehabilitation Act was part of the new program.

The radio broadcasts, which Bennett himself paid for, modelled after President Franklin Roosevelt's successful use of radio, were a first for Canada. They were carried by thirty-nine stations across the country, and by the time of the fifth broadcast it has been estimated that some 8 million Canadians were listening,[2] a remarkable claim with the country's population then under 11 million.

Not many of those listeners were in Saskatchewan. There were only about 20,000 radios in the entire province, and few of them were in not-yet-electrified rural areas, where battery-operated sets were expensive.[3] (Newspaper readership also suffered during the Depression. In Regina, the subscription lists of the *Leader-Post* fell from 130,000 to 35,000 and of the *Regina Daily Star* from 70,000 to 14,000.[4])

In February, Minister of Education Bill Estey presented to the legislature the results of the Debt Survey on the level of public debt in the province. The situation turned out to be not as bad as feared.

Rural municipalities generally were in good shape financially, and even the sixteen RMs in the severe drought area were in surplus, but only if their indebtedness for relief was removed. Villages and towns were in similar conditions, though quite a bit of property had been taken back for taxes. Rural schools had suffered severely when taxes dried up but had slashed expenditures (e.g., teachers' salaries) to compensate and thus carried little debt.

All in all, the Debt Survey went a long way toward dispelling the view that Saskatchewan was a bankrupt province and enabling some confidence to be restored.

When the spring of 1935 arrived on the prairies, it brought rain and hope that the years of drought had finally lifted. The crops started out with great promise. In his diary, Charles reported strong rains all over the south in May and June and into July. The fields held *nice crops*.

But no one was under any illusion that the Depression was ending. During the last half of June, Regina was flooded with some 2,000 unemployed men heading to Ottawa to protest their condition. Prime Minister Bennett ordered that the On-to-Ottawa Trek, as it became known, be stopped at Regina. The result was the Regina Riot on Dominion Day, July 1st, and two deaths, one trekker and one city policeman.

Charles had been in Regina earlier that unfortunate day and noted the arrival of another calamity. The wheat fields began to show signs of stem rust, a fungus whose spores had been carried in by the winds out of the Dakotas. *Bad rust in Regina* his July 1st entry reported. As his summer travels extended, there were more such accounts. July 24th, Radville to Estevan: *Rust serious.* July 25th, Lampman, Kisbey, Kenosee, Fort Qu'Appelle: *Lots of bad rust.* August 3rd, Cupar: *Country very wet. Rust bad.* August 5th, Wolseley: *Saw bad rust.*

George Britnell confirmed Charles' observations:

In 1935 confident anticipation, following a break in the drought, of at least average wheat yields were shattered when rust invaded the province from the east, drought from the west, and frost from the north. The extremely severe epidemic of wheat-stem rust completely destroyed nearly one-quarter of the total wheat acreage and materially reduced both the quality and quantity of much of the remaining crop.[5]

Another pestilence, another crop gone. In addition to drought, soil drifting, grasshoppers, and now wheat-stem rust, there was one more plague quietly consuming much of such crops as the ever-suffering farmers could coax from the soil. The Richardson ground squirrel, commonly known as the gopher, or flicker tail, had been present on the plains long before the arrival of settlers. A small (ten to twelve inches at maturity) burrow-dwelling rodent, it feeds on plants, usually the leaves, seeds, and roots of grasses. The planting of cereal grains where buffalo grass once grew was like manna to the resident gophers. They converted their taste preferences and prospered, and verily their tribe did increase.

Complaints from settlers that their crops were consumed by gophers were recorded as early as 1884 when the famous Bell Farm at Indian Head offered a half-cent bounty for gopher tails. In 1889, "a special committee of the Northwest territorial assembly determined that certain districts had suffered 52 per cent losses of crops to gophers."[6]

In Saskatchewan, by 1907, the newly formed Department of Agriculture offered financial incentives to municipalities that paid bounties on gopher tails. In 1916, it designated May 1st as provincial Gopher Day and encouraged killing the rodents, particularly by children. Prizes were awarded to those with the highest body count, and the reported mortality totals responded. By 1921, the last year of the program, six Shetland ponies headed the prize list as more than 2 million gopher tails were turned in.[7]

Children employed a number of methods of gopher extermination: trapping them, snaring them, drowning them out of their burrows with buckets of water, or shooting them with the ubiquitous .22 rifle. The last method had two drawbacks: ammunition was costly, and even a mortally wounded gopher would strive to escape down its hole, which would save, if not its life, at least its tail.

Bruce Peel, in his monumental study of the Rural Municipality of Mankota, writes that poisoning gophers began there as early as 1914, when May 1st and 2nd were declared special gopher extermination days, and all were encouraged to poison gophers. The RM also had a bounty program and in 1920 paid out for 70,530 tails. One boy (Alton Headrick, who later farmed at Limerick) brought in 5,805 tails.[8]

In the 1930s, gophers became even more of a menace, perhaps because, as was widely believed, they thrived in times of drought, or perhaps because the crops were sparse and the rodents and their depredations more noticeable, or both. Certainly the gopher became a real and continuing menace all through the Depression years.

Poison became the preferred method of extermination, with strychnine in common use. But, to be effective, it had to be disguised with flavouring to cover its bitterness. The Department of Agriculture distributed instructions for mixing strychnine bait, and several privately invented concoctions were promoted.

One of the most successful was Kay's Killer, a recipe devised by Scott MacKay, the Limerick druggist whose drugstore was just across Main Street from Wilson Brothers. Kay's Killer was also strychnine based, mixed with corn starch and then licorice, which overcame the bitter taste. The product was mixed again with bran for application.

The demand for Kay's Killer turned its early modest production into a cottage industry. MacKay and Joseph Smith, a local farmer with a unique talent for things mechanical, took over an abandoned garage and created a seasonal assembly line that employed all the MacKay children canning and labelling the product. Kay's Killer was distributed all over southern Saskatchewan.

Prime Minister Bennett's bold new program finally went before voters for approval on October 14, 1935. The omens were not good. In Bennett's home province of Alberta, an election on August 22nd had defeated every candidate that the governing United Farmers had put forward and, out of the total of sixty-three seats, elected fifty-six

MLAS from the new Social Credit Party under William (Bible Bill) Aberhart. It has been called the largest single electoral shift in North American history.

Charles was in Regina that election night and, with a couple of friends, a radio, and a bottle of scotch, settled into a room in the King's Hotel to listen to the Alberta results. The Liberal Party hoped to profit from a scandal that had badly wounded the United Farmers government. When the avalanche was over, the astounded group noticed that they had been so absorbed in the cataclysmic radio reports that the scotch had been forgotten.

In the election called for October 14, 1935, Prime Minister Bennett and his Conservatives fared somewhat better, but they lost almost 100 seats in the 245-seat Parliament, dropping from 134 to thirty-nine, while the Mackenzie King Liberals won 173. In Saskatchewan, sixteen of the twenty-one seats went Liberal, two CCF, two Social Credit, and one Conservative. Charles' MP, Dr. Donnelly, won again, this time in the new riding of Wood Mountain.

Returned to the prime minister's office, Mackenzie King dashed Dr. Donnelly's hopes and instead recruited Premier Gardiner to his new cabinet, and on October 28, 1935, Gardiner was sworn in as Canada's minister of agriculture, a position that he would hold for twenty-two years. He secured election to Parliament from the constituency of Assiniboia in a by-election held on January 6, 1936.

The Liberal caucus met in Regina on November 1st to choose one of their members to succeed to the premier's office. The Liberal Party Council, which included Charles, also met that day. The party's approval of the caucus decision would be needed.

Only two ministers put their names forward, William (Billy) Patterson, minister of telephones, and Thomas (Tommy) Davis, attorney general. The vote was close, so close that when Davis, who did not crave the position, voted for Patterson instead of himself it turned out to be the winning vote.

Caucus then joined the Liberal Party Council and reported their decision for Patterson. Charles, who knew both men, but neither well at that time, promptly moved that the council ratify the caucus vote, and it was done. It was a decision that Charles regretted when, a few years later, he worked with Davis and came to know his abilities, and they developed a great friendship.

In my own memories I count this as the most serious political mistake that I made in my lifetime. Years later in Canada House in London I reminded Tommy Davis of it and expressed the same opinion. His reply was characteristic. "Well, if you had given it to me, it would not have died in the ditch."

But Davis did not admit to Charles that he had not really wanted to win.

There were ample reasons why an intelligent man would not be attracted to the premiership of Saskatchewan in 1935. The results of the Debt Survey notwithstanding, the province's finances were in dreadful shape following four consecutive years when net farm income had fallen into the red. And Saskatchewan pretty much stood alone in its fiscal difficulty. No equalization formula had yet been developed to ease the burden of low-revenue provinces.

At Limerick, Charles managed to keep Wilson Brothers running but with difficulty. His son Kevin had joined the office in 1933 after completing senior matriculation and then Success Business College in Regina. Charles had the view, very prescient for the time, that Canada was a bilingual country and that his children would be better equipped if they spoke French. Accordingly, Kevin spent two years at College Mathieu in Gravelbourg, where the instruction was entirely in French. Daughter Moira, who had graduated from grade twelve at sixteen, was spending the 1935–36 year at the Daughters of the Cross Convent in LaFleche before going on to university. Both became quite fluent in French.

Sending the two eldest Wilson children to French Catholic institutions was not a popular move in Limerick, a heavily Protestant community with a strong Orange Order lodge.

By the spring of 1936, it was becoming apparent that a solution to the farm debt problem could not be awaited much longer. Farmers were getting really restive. After all, practically every one of them was in arrears on his mortgage, and the mortgage was on his home. I became aware that many meetings were being held Sunday afternoons in country schoolhouses and that these meetings were being infiltrated by agitators who were expressing opinions and suggesting remedies which nobody really wanted.

To get to the root of this matter, I called a meeting, a general meeting, of the Liberals of three provincial constituencies [Willow Bunch, Bengough, and Moose Jaw County] centering on the Town of Assiniboia. I invited the new premier of the province, the Honourable W.J. Patterson, and his attorney general, the Honourable T.C. Davis, who were the two members of the government most acutely concerned with the debt problem. Both gentlemen were warned that they should stay away from the meeting, that they would probably end up by having their clothes torn off them. But I assured them that they would meet with nothing but respect and attention at the meeting, and I proved to be right.

The meeting was held in the skating rink at Assiniboia, a metal-clad building, on June 24th, a hot afternoon, and it was a most uncomfortable meeting to attend as the temperature in the rink must have been in the neighbourhood of ninety to 100 [degrees Fahrenheit]. That really proved an asset to me because it made the audience tired sooner than it might otherwise have become so.

On the platform with me were the premier and the attorney general, forewarned by me that they were not to speak, that it was their duty to listen all afternoon so that they would be well apprised of the ... feeling in the country. The meeting was attended by some 1,600 farmers, and the debate was keen and active. Countless suggestions came from all corners of the hall, most of them utterly impractical. But we had set up a resolutions committee two weeks in advance, of which I was chairman, and we had prepared what we hoped would be an acceptable resolution, both to the government and to the farmers.

Along in the late afternoon, I complimented the meeting upon the acute attention they were giving to the problem, and the spirited debate which had taken place, and told them we had prepared a resolution which we hoped would embody their sentiments, and I thought the time had now come to present it to them. I then had the resolution read to them by a Liberal who had a voice like the Bull of Bashan and who made sure it was heard in every corner of the hall. To my delight, it was accepted.

The resolution was a lengthy recital of the severity of the farm debt problem, a claim that such debt adjustment as had taken place in the past two years had not even touched the fringe of the problem, and then urged the government to do far more to rehabilitate the agricultural sector.

The time was now about 5:30. I again complimented the audience on their worthwhile labours during the afternoon and suggested that many of them were thinking of a dinner interval. I said, "I am going to give you a long one. I think we should adjourn right now so that you can have two and a half hours to rest and relax, but I want you back here at eight o'clock to listen to the two members of the government who are most concerned with your problem."

That program was carried out, and a nice, quiet, respectful meeting resumed at eight o'clock and listened to addresses by Mr. Patterson and Mr. Davis, no doubt tailored quite a bit to suit the occasion.

At the conclusion of the meeting, a French farmer came to me and said, "Charlie, you have written history here this afternoon. Never before in this province has the government sat on the platform and listened to the people, but you accomplished that here today, and nothing but good can come of it."

The Assiniboia meeting, and the resolution there enacted, were covered some weeks later by the *Western Producer*, a farm weekly published in Saskatoon with a huge readership. The lead paragraph of the story spoke of how the meeting viewed the desperate situation in the southern drought area.

That debts of Southern Saskatchewan farmers can never be paid is the considered opinion of Liberal party supporters from the provincial southern constituencies, who point out that crop failures and world conditions in the past few years have reduced the population in what was once Canada's finest wheat growing area to a point where, even if they are forced to accept "slave standards of living" and perpetual tenancy in place of ownership of their farms, there is little hope of their obligations being discharged within the lifetime of the present generation, or at all.[9]

As to the resolution, which was set out in full, the *Western Producer* stated in an editorial that "This is a moderate and apparently carefully thought-out statement of the case. Whether one agrees fully with it or not, it reflects credit on those responsible for drafting it."[10]

The times were desperate, and Charles was not unaffected. He was struggling to keep Wilson Brothers afloat, and meeting its modest payroll was a constant effort. As the Assiniboia meeting showed, he was a Liberal of some stature. With Liberal governments in both Regina and Ottawa, he could easily have called in some IOUs and secured a government position but chose not to.

Florence and I had decided that we would not accept any government job whatever, that we would stand on our own feet through the Depression and have a little business to carry on with when the storm was over. We had served the Liberal Party so diligently throughout the years that the offer of a good government job could be expected, and as a matter of fact we were offered several but declined them. I was offered both the federal and the provincial nomination in this seat more than once but had the excellent sense to remain aloof from them too. We simply went on serving the party if, as, and when an opportunity occurred and frequently going outside that and creating the opportunity. But I decided that I could serve the party better from the office at Limerick than I could by occupying one of the seats, and so we remained here.

But the inexorable law of politics as practised in the 1930s would dictate a different destiny for Charles. At the time of the 1935 federal election, some 273 people were working on debt adjustment under the Farmers' Creditors Arrangement Act.[11] Such a source of employment was considered legitimate spoils due to the victorious Liberal Party, and practically all, if not all, of the positions were vacated, and the recruitment of Liberal successors began.

ENDNOTES

1. Cited in Ernest Watkins, *R.B. Bennett: A Biography* (Toronto: Kingswood House, 1963), 257.
2. Ibid., 220.
3. Pierre Berton, *The Great Depression 1929–1939* (Toronto: McClelland and Stewart, 1990), 283.
4. Wayne Smaltz, *On Air: Radio in Saskatchewan: 1990* (Regina: Coteau Books, n.d.).
5. Britnell, "Saskatchewan, 1930–1935," 51, 52.
6. Thomas D. Isern, "Gopher Tales: A Study in Western Canadian Pest Control," *Agriculture History Review* 36, 2 (1988): 189.
7. Ibid., 193.
8. Peel, "R.M. 45," 231–34.
9. *Western Producer*, September 3, 1936, 7.
10. Ibid, 3.
11. W.T. Easterbrook and W.B.H. Easterbrook, "Agricultural Debt Adjustment," *Canadian Journal of Economics and Political Science* 2, 3 (1936): 392.

CHAPTER 17

THE BOARD OF REVIEW

In Ottawa in early 1936, members of the Saskatchewan Liberal caucus, including the newly appointed minister of ariculture, James Gardiner, met to determine who would be appointed as members of their province's Board of Review. Chief Justice J.F. Brown of the Court of King's Bench was selected to be chairman, R.J. Moffat of Bradwell to be creditors' representative, and Charles Wilson of Limerick to be farmers' representative.[1] Charles had not been asked if he would agree to serve and did not know of his selection until word seeped back to him.

In 1935, everyone knew that there would be a change of personnel in the Saskatchewan Board of Review. The previous board had got off on the wrong foot the very first day they opened for business due to the fact that none of them had any really intimate knowledge of the terrific situation in respect of debt which had grown up here. I had been on the road allowances steadily and had accumulated a wide and intimate knowledge of the situation. And, moreover, from the very outset of the storm my sympathies had been with the farmers on the land. Indeed, from the point of view of the interests of the loan companies whom I represented, I regarded it as essential that the settlers should be retained on the land. I could see no value in this land if it did not have people on it. I had taken a prominent part in the discussion of the problem, not only locally but at widely scattered points across the province. The result of this was that, when the matter of an appointment of a new farmer commissioner to the Board of Review came up at Ottawa, my name was instantly mentioned and was the only one mentioned, so that the position was in my lap.

At the outset, I shied away from it in pursuance of the policy which Florence and I had agreed upon in our home, although God knows we needed the job very badly indeed. It seemed to be common knowledge across the province that I was to be the new farmer member to the board.

The proposed appointment of Charles was not the only one becoming known. That of Moffat as creditors' representative was causing some difficulty. The representative on the previous board was Gordon Forbes, a Conservative, of the Regina law firm Cross, Jonah, Hugg and Forbes, a firm that represented a large number of the loan companies whose mortgages were being considered by the Board of Review. The loan companies were not satisfied with the proposal that Moffat represent their interests. Over the objection of the Saskatchewan Liberal caucus,[2] Robert Hugg, of the same law firm, but a Liberal, became the intended appointee.

On July 21, 1936, Charles was in the midst of his annual inspection tour of the many mortgages that he supervised; he was staying at the Assiniboia Club, the prominent men's club in Regina where he had been a non-resident member since 1918. That morning he ran into his friend Archibald Peter McNab, a colourful politician known to all as Archie. (*I met Archie in the early days, trading grain and sitting on the sidewalk in his bare feet.*) McNab, an MLA from 1908 to 1927 and minister of public works from 1912 to 1926, was hoping to be appointed shortly as Saskatchewan lieutenant governor.[3]

Archie was discussing my appointment and was shocked to learn that I did not propose to accept it. During the conversation, I said to him, "I have to make a trip to Wolseley this afternoon. You come along for the ride and keep working on me. Perhaps when I get back this evening you will have changed my mind."

He came, and just as we were walking out of the club I was called to the phone, and Miss Allonby in my office at Limerick read me the telegram from Ottawa offering me the appointment. I went out to the car and so informed Mr. McNab, who tried to get me to reply to it immediately, but I stalled him, and we made the trip to Wolseley together.

It was a little late in the evening when I returned to the club, and I went into the dining room and had dinner all alone, still thinking furiously. When I left the dining room, I happened to run into Percy Hodges, then a very prominent lawyer in Regina, a great Liberal, and a great friend of mine.

"What are you going to do with that telegram in your pocket?" Percy asked me.

I put on the appearance of coyness and replied, "What telegram?"

"You do not need to be so coy about it. We all know about it. The point is what are you going to do with it?"

I told him I was just on my way to the writing room to indite a reply declining to serve. He immediately became concerned and told me, "Charlie, you can't do that, for three reasons, as I'm going to place in front of you.

"The first one is you have made yourself prominent at the discussion of this question all across the province, and you have led people to believe that you know how to solve it. It has to be solved for the sake of the Liberal Party, and most people think you are the man to do it.

"The second reason is that, because you have been so prominent in discussing the matter, you have led people to expect that you know the answers. Now you have an opportunity to apply them, and, if you flunk us, people will say 'He was glib enough at prescribing remedies, but when the opportunity came to him to apply his remedies, he did not have the courage.'

"The third reason is a personal one. Unless you are in better financial shape than we around the club think you are, you need the job."

I replied, "Percy, to deal with the third reason first, I am bankrupt, as nearly everybody is, and I have never tried to conceal it around the club. But that in my mind is a minor consideration in this matter. The other two of your arguments have great weight, and you are obliging me to think the matter over again."

He immediately linked his arm in mine and said, "Come on to the writing room, but I will write the telegram, and you will sign it."

And that is the exact manner in which I arrived on the Board of Review.

The formal appointment of Charles to the Saskatchewan Board of Review was made in the style of the time, under the great seal of the governor general, signed by the chief justice of the Supreme Court of Canada, Lyman Duff, as deputy governor general, on August 17, 1936.

My conscience was still greatly disturbed because I did not think that any judge, Liberal or Tory, would allow the Board of Review to go to the extent in reducing debt which I knew they had to go if the problem was going to disappear.

The day the previous board opened, the chairman announced that Parliament had endowed the board with the authority for cancelling debts, but he added that it is an authority which we will use very sparingly. That stultified the board and disappointed and frustrated the farmers who knew as well as I knew that the word sparingly did not belong, that what the debt needed was an axe in the hands of a vigorous man.

One afternoon I was in Regina, and Charlie Hoffman, the registrar of the Board of Review, called me on the phone and told me that Chief Justice Brown, the new chairman of the board, would like to have an interview with me. I made myself available immediately and spent a couple of hours with that wonderful gentleman, in whom the farmers of this province never had a better friend. When that interview was over, I had the conviction that the new board could really go places under his chairmanship, and my mind was a little more at rest.

The appointment of Charles as farmer, or debtor, representative on the board presented something of a problem. Undeniably he was, and had been for years, a loan company man. How could he now be expected to serve contrary to the interests of his former employers?

When I joined the Board of Review in August of 1936, I immediately resigned from the services of Victoria Trust and Savings Company.

However, they refused to accept the resignation. They told me they were familiar with my views on debt adjustment, and they were prepared to go along with them.

They asked me to set up at Limerick some kind of a temporary organization which could function until I got through with the board. I was very happy to do this because it gave me a prospect of having a little business to return to after the activities of the Board of Review came to an end.

My son Kevin was then twenty years of age, and I placed him in charge of the Limerick office. By way of assistance, I secured the services of my old friend Roscoe Hiram Pierce, who had quite a little experience in the field closely allied to the mortgage field.

The Saskatchewan Wheat Pool accepted the news of my appointment with, let me say, very quiet reservation. I had been all my life connected with the loan companies of the province, and now I was the farmer representative on a very important board from their point of view. Their attitude was natural, even though their board contained my very good friend and brother-in-law, A.F. Sproule. But he shared their general conviction that, although my compassion was with the farmers, I had worn the livery of the loan companies for too long a period to shed it very easily.

I met this situation at a very early date by seeking an interview with the Wheat Pool board and telling them frankly that I was aware of their attitude. I asked them to do just one thing for me—hold their opinion in suspense for six months and let us see how it worked out. They agreed to do that, and I want to say that from that point on I had no better friends or colleagues in the Province of Saskatchewan than the Wheat Pool and the members thereof.

At the time we took office, we had a large room in the Westman Chambers building in Regina. I immediately made it my practice to sit on the opposite side of the table from the judge and the creditor representative because I had a life-long religion for looking a man in the eye when I was speaking seriously to him.

The new creditor representative was Robert W. Hugg, KC, a Regina lawyer who in his office represented most of the loan companies of the province. He was a gentleman of outstanding ability and unassailable integrity, but he was very obstinate and persistent in the views he held.

The Saskatchewan Board of Review in action at the Saskatoon Court House in 1939: Justice T.C. Davis, chairman, flanked by Robert W. Hugg, KC, creditors' representative, and Charles Wilson (partially obscured), farmers' representative. In the background are C.T. Olding, KC (right), North Battleford, board solicitor, and Dan A. Sneaker, Regina, board secretary.

StarPhoenix photo, Saskatchewan Archives Board.

The first morning we got together Chief Justice Brown produced a statement which he proposed to hand to the press dealing with the mortgage rate of interest. He had inserted 6 percent without consulting us. When the statement was read in the boardroom, Mr. Hugg firmly advised that he adhered to a rate of 7 percent.

Mr. Hugg then pointed out that the situation made mine the deciding vote, and he added that he wished to warn me before I cast it. He then told me that some of the loan companies could not possibly meet the 6 percent rate and that, if I voted for it, the immediate effect would be that at least two of them would be forced to close their doors, and he wanted me to understand the responsibility that rested on my shoulders.

I knew enough about the general situation to be well aware that forcing two of these companies to close their doors would be no real contribution, and I turned and walked the length of the boardroom three or four times before coming back to my place. Then I told him, "Hr. Hugg, I will in turn tell you what will happen tomorrow morning if I vote for 6 percent. There will not a damn thing happen, and I do vote for 6 percent."

And so 6 percent became from that time on the standard rate of interest on all mortgages that came before us in place of the 8, 9, and even 10 percent which had prevailed previously.

Our board met for the first time publicly at the courthouse in Saskatoon in August of 1936. We received a lot of briefs dealing with the debt situation. Among them was one presented by G.R. Bickerton, who was then president of the Saskatchewan Section of the United Farmers of Canada. I did not know Mr. Bickerton, and I suggested to Chief Justice Brown that it would be a good idea for me to get acquainted with him. The chief approved, and I followed Mr. Bickerton into the corridor and invited him to lunch with me to give us an opportunity to get acquainted. Mr. Bickerton came, and we had a full discussion, and a frank discussion, and after that I never had any difficulty getting along with the United Farmers. I frequently invited the opinion of Mr. Bickerton on the policies we were pursuing, and I found him at every turn of the road a loyal friend and a wise counsellor to whom I am still greatly indebted.

In July 1936, Minister of Agriculture James Gardiner and his provincial counterpart, Gordon Taggart, did an extended tour of the western drought area. They found little to encourage them. By then, almost 14,000 farms had been abandoned, 8,200 in Saskatchewan, 5,000 in Alberta, and 700 in Manitoba.[4] The movement away from the land continued almost daily, and the forsaken soil followed on the ceaseless winds.

The summer of 1936 was not a time for tourism in southern Saskatchewan. A record heat wave scorched the prairies, and much of Canada, for more than two months. Beginning in June, the furnace-like winds blasted the crops in their fields, taking the drought for the first time into northern Saskatchewan.

By July, Regina residents were bedding themselves down in the city's parks at night in futile attempts to escape the oppressive heat. For a full week, the Fahrenheit thermometers registered more than

100 degrees from Calgary to Toronto. Hundreds died from heat prostration.

In the west, the unrelenting heat wave continued well into August. The Saskatchewan crop yield dropped to its lowest level yet since the Depression began, a paltry 7.5 bushels per acre for the entire province.[5] In the southern portion, many fields did not return seed.

Attempts to quantify the total agricultural debt of Saskatchewan were nearly futile as the target rose rapidly and relentlessly. The best estimate was that the debt stood at $525 million at the end of 1936.[6] Only 5 percent of Saskatchewan farms were believed to be debt free, and likely none of them was in the southern section. Another $200 million was needed to restore the rundown Saskatchewan farms and equipment to operating capacity.[7]

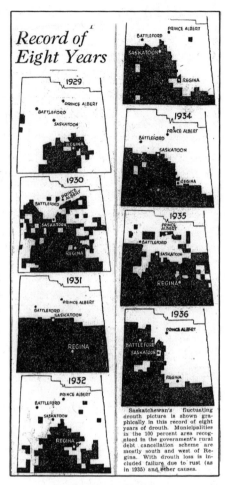

Drought map, Regina *Leader-Post*, September 29, 1936.

A valid concern was whether the slow-moving machinery of the Farmers' Creditors Arrangement Act, supplemented by the provincial Debt Adjustment Board, could cope with this ever-increasing mountain of debt. Gardiner took immediate command of his agriculture portfolio and vigorously attacked the escalating problems in the west. Unfortunately, around the cabinet table in Ottawa, he found himself pitted against ministers representing central Canada with competing concerns for their own regions. Prime Minister Mackenzie King gave

Gardiner only lukewarm support. Although Saskatchewan supported him with a strong representation of sixteen out of twenty-one MPs, more than any province other than Quebec and Ontario, King had a despairing opinion of the region. "It is part of the U.S. desert area. I doubt if it will be of any real use again," he confided to his diary.[8]

Some part of the image problem suffered by the west, Saskatchewan in particular, stemmed from the Social Credit government in Alberta and its often maverick attempts to deal with the problems of the Depression. Premier Aberhart was in a constant battle with Ottawa and the Supreme Court over his unorthodox fiscal legislation. Attorney General of Saskatchewan T.C. Davis described the situation well in a letter to Gardiner on August 28, 1936: "This Province of Saskatchewan is the sheet anchor for sanity in Western Canada, and, therefore, the sheet anchor of Canada as a whole."[9]

Gardiner was determined to preserve western agriculture and the western farm population. At his initiative, the mortgage and loan companies were induced to enter a major farm debt reduction agreement with the federal and provincial governments.

In late September 1936, Premier W.J. Patterson announced the agreement. The Drought Area Debt Reduction Plan, as it was called, provided for some $75 million of debt cancellation, one-third absorbed by the mortgage companies, another third in tax relief, and the last third in relief and seed grain advances. In addition, the interest rate on all farm mortgages in Saskatchewan, not just those that came before the Board of Review, was reduced to 6 percent effective January 1, 1937.

The cancellation was directed to the drought area, defined as 158 rural municipalities and local improvement districts in southern and western Saskatchewan. The plan was promptly implemented by legislation.[10]

Within a few days, cynical minds pointed out that the actual amount being written off by the loan companies was less than $15 million, almost all of which was totally uncollectable in any event, and that cancelling taxes and government relief liens that had priority over mortgages merely served to improve the positions of loan companies.

The best method of attacking Saskatchewan's debt problem remained as it had been in the beginning—a few good crops accompanied by decent prices. But any hope of relief from nature faded as the scorching heat of the summer of 1936 turned into a rainless fall and then into an almost snowless winter, with vicious dust storms that turned the few meagre drifts black. By the spring of 1937, it was apparent to all that the worst year of the worst of times had descended on the already parched and suffering plains.

ENDNOTES

1. LAC, Mackenzie King Papers, vol. 219, 188211, J. Fred Johnston to W.L. Mackenzie King, August 4, 1936.
2. Ibid.
3. McNab was so appointed on September 10, 1936, and served with flair until February 26, 1945.
4. Gregory P. Marchildon, "The Prairie Farm Rehabilitation Administration: Climate Change and Federal–Provincial Relations during the Great Depression," *Canadian Historical Review* 90 (June 2, 2009): 293.
5. Britnell, *The Wheat Economy*, 50.
6. Ibid., 87.
7. Britnell, "Saskatchewan 1930–1935," 163, 162.
8. Cited in Robert A. Wardhaugh, *Mackenzie King and the Prairie West* (Toronto: University of Toronto Press, 2000), 206.
9. Cited in Norman Ward and David Smith, *Jimmy Gardiner, Relentless Liberal* (Toronto: University of Toronto Press, 1990), 206.
10. Drought Area Debt Adjustment Act, 1937, Statutes of Saskatchewan, c. 92.

CHAPTER 18

1937: THE YEAR FROM HELL

In the spring of 1937, it was as if the gods of nature had tired of the slow and agonizing torture inflicted on the prairie west for eight years and determined to administer a *coup de grâce*. Precipitation almost disappeared. At the Assiniboia recording station, roughly in the centre of the southern Saskatchewan region, the total for April, including a little snow, was 9.7 millimetres, or .40 of an inch. For May, 12.7 millimetres, or .51 of an inch. For June, 17.3 millimetres, or .66 of an inch.

Not enough to settle the dust, much less germinate a crop. And the temperatures and winds ensured almost immediate evaporation of what little moisture there was. In April, thermometers rose to 65 degrees Fahrenheit (18.35 Celsius), in May to 84 (28.95), in June to 102 (38.9), and in July to 109 (42.8). On July 5th, the all-time Canadian record for heat was set at Yellow Grass and Midale: 113 (45).

Sloughs and potholes had already become memories, but now even lakes disappeared. The ceaseless winds tore the alkali from the dried-out bed of Old Wives Lake near Mossbank, whitening the surrounding countryside.[1] The same winds ripped soil from exposed fields, creating towering and frightening dust storms that in turn left drifts that blocked roads all over the south country. The term "blowdirt" entered the western lexicon.

Southern Saskatchewan and Alberta suffered grievously from
dust storms during the 1930s, but the real horror was endured by
residents of the American Great Plains, the vast, treeless, original
prairie region that stretched 4,000 kilometres (2,500 miles) from
western Canada down to Mexico. Averaging 650 kilometres (400
miles) wide, it encompassed almost all of Kansas, Nebraska, North
and South Dakota, most of Montana and Oklahoma, as well as good
parts of Texas, Colorado, Wyoming, and New Mexico.

The lack of trees and the aridity caused early explorers to name
the region the Great American Desert, a description that became apt
during the Dirty Thirties. The Great Plains, or Desert, extended into
western Canada as far as the Saskatchewan River.

The greatest cruelty fell on the southern portion of the Great Plains,
particularly the Texas and Oklahoma panhandles, New Mexico, and
Kansas. There in 1933 multiple dust storms lasted seventy days. The
one bad year of 1937 saw 134 storms, while 1938 endured a storm that
lasted three days. There was so much dust in the air that people,
mostly children, developed "dust pneumonia," a form of silicosis that
frequently brought death by suffocation.[2]

On May 9, 1934, a dust storm that probably originated in Alberta
and Saskatchewan blew out of the Dakotas and eastern Montana. It
is described in *The Worst Hard Times:*

> The next day, a mass of dust-filled clouds marched east,
> picking up strength as they found the jet stream winds, moving
> toward the population centers. By the time this black front hit
> Illinois and Ohio, the formations had merged into what looked
> to pilots like a solid block of airborne dirt. Planes had to fly fif-
> teen thousand feet to get above it, and when they finally topped
> out at their ceiling, the pilots described the storm in apocalyptic
> terms. Carrying three tons of dust for every American alive,
> the formation moved over the Midwest. It covered Chicago at
> night, dumping an estimated six thousand tons, the dirt slink-
> ing down walls as if every home and office had sprung a leak.
> By morning the dust fell like snow over Boston and Scranton,
> and then New York slipped under partial darkness. Now the

storm was measured at 1,800 miles wide, a great rectangle of dust from the Great Plains to the Atlantic, weighing 350 million tons. In Manhattan, the streetlights came on at midday and cars used their headlights to drive. ... On May 11 the orphaned land of the Great Plains came to the doorstep of the nation's premier city. For five hours, the cloud dumped dirt over New York. Commerce came to a standstill. ... The storm moved out to sea, covering ships that were more than two hundred miles from shore.[3]

The ceaseless winds tore the topsoil from 33 million acres of American farmland. In Canada, the soil was often stripped down to the hardpan, sometimes as deep as three feet. The heart of the North American continent was ripped away.

In Saskatchewan during that spring of 1937, nothing grew. No crops, not a blade of grass in the pastures, not even weeds, not even the tenacious Russian thistle, the tumble-weed that was the ever-present symbol of the Dirty Thirties, and 1937 became the year that epitomized the Dirty Thirties. To those who endured it, 1937 was the Year from Hell.

There was no feed available for the tremendous livestock population. Most farms were still operated by horse power, and after grain prices had dropped through the floor cattle numbers had increased as farmers turned more to mixed farming for survival. But cattle prices fell also, and instances of cattle being shipped to market in Winnipeg and not bringing enough to pay the freight bill were not uncommon. Naturally enough, the animals remained on the farms wherever possible. The Saskatchewan farm livestock population had increased from 1,189,000 animals in 1931 to 1,535,000 in 1936 and then dropped slightly to 1,441,000 by 1937, still far too many. With no feed available anywhere, the federal government took strong steps to remove hundreds of thousands of head before they starved to death and managed to get the population down to 1,129,000.[4]

The program provided for payment of one cent a pound for animals shipped and another one cent to packers for processing them. The freight costs to Winnipeg were picked up, so the delivery price became the on-farm price. Arbitrary limits were placed on the number of animals that a farm was allowed to maintain. A huge cattle-marshalling facility was constructed at Carberry, Manitoba, where the animals were sorted and auctioned.

Then, in the last half of July, the rains returned. By the end of the month, Assiniboia had recorded 65.8 millimetres, or 2.6 inches; August brought another 28.7 millimetes, or 1.1 inches, and September 63.8 millimetres, or 2.5 inches. Not huge amounts but more than enough to germinate the Russian thistle that for eight years had been rolling around the west depositing seeds. The weed luxuriated in near-perfect conditions, and farmers rushed into fields to capture the God-sent hay. The product was frequently described as "About as nutritious as straw and as tasty as barbed wire," but, with the bit of feed grain that was available, it was just enough to carry animals over the winter.

But there was no crop and thus no income. The average wheat yield for the entire province fell to a miserable 2.7 bushels per acre, one-third of the 1936 return, and the lowest since meaningful record keeping began in 1900. In most of the southern region of Saskatchewan, the return was zero. The provincial Department of Agriculture's statistics employed the word *nil* to identify the cash return per acre of five crop districts, Regina–Weyburn to the Alberta border.[5] Saskatchewan's net farm revenue fell into the red again, this time to a loss of $36,336,000, the worst of the five red-ink years of the 1930s.

The efforts of the official receivers and Board of Review under the FCAA, the provincial Debt Adjustment Board, plus the $75 million reduction granted by the Drought Area Debt Adjustment Plan, and other concessions, made little impact on the huge farm debt. Estimated at $525 million at the end of 1936, the total for the end of 1937, after adding back another year's interest and taxes, was still a formidable $475 million.[6] And it was headed back up.

Chief Justice Brown's Board of Review, with Charles and Robert Hugg as commissioners, was quick off the mark and worked industriously against the swelling tide of farm debt, but the early results, though encouraging in individual cases, had almost no impact on the overall problem. If each farmer who came before them seeking help was to receive careful consideration of his affairs, the board found that ten to twelve cases were all that could be handled in a day. The number of Saskatchewan farmers in economic distress was estimated to be well over 100,000.

We were making only slow headway, and the United Farmers kept reminding us that we had an appalling amount of work in front of us. I remember Mr. Bickerton telling me that there were 140,000 farmers in the province, and, in his opinion, 110,000 of them needed surgery.

We were able to deal finally with about 1,100 cases a year. Measuring that against what was alleged to be ahead of us made it look like a task the end of which would never be reached.

The criticism that Charles directed at the earlier board, serving from 1934 to 1936, was at least somewhat well founded. When the cases that it had dealt with were analyzed, it was found to have directed reductions in farm debt averaging 27.7 percent and cuts of annual interest per farm of $171. By 1937, the Brown–Wilson–Hugg board had achieved reductions averaging 45.5 percent and cuts of annual interest amounting to $371.[7] That higher level of debt reduction was to continue throughout the existence of the board.

The process that a farmer had to endure to take his affairs before the Board of Review was slow and somewhat cumbersome. He began with a visit to the official receiver at the courthouse in his judicial district. In the beginning, this was often the sheriff, who performed a similar function for the provincial Debt Adjustment Board. As the number of desperate farmers taking advantage of the procedure increased, so did the number of official receivers, until there were thirty-seven operating in Saskatchewan's twenty-one judicial districts.

The farmer's financial condition was carefully laid out, with all debts itemized, a list usually headed by the farm mortgage, followed by implement dealers, merchants, and, almost always, an amount for relief received and repayable. Then attention was paid to assets, the land at its current value, machinery, livestock, and incidentals. Next came a short history of the farm's operation, from which an assessment of the land's productivity and the farmer's skill and management ability could be made. As well, attention was paid to the farmer's family and their needs.

Then the official receiver notified all the creditors and called them to a meeting with the farmer to see whether a proposal could be worked out and agreed upon. In most cases, such a proposal involved cutting the liabilities down to, or close to, the current value of the assets. Usually only the larger creditors, such as the mortgage loan company or bank, would show up at this meeting.

If an arrangement could be worked out, the official receiver was authorized to put it into effect. If no agreement could be reached, either the farmer or a creditor could apply to have the file referred up to the Board of Review for consideration at its next sitting, which in that judicial district could be several months away since the board toured through the province's twenty-one districts. But the board would not accept cases where the amount owing between the farmer and a creditor was in dispute. Those disagreements had to be referred to the conventional courts for resolution. As the policy of the board became better known, and its likely decision in a given case more predictable, far more settlements took place at the official receiver level.

In March 1937, the Board of Review was sitting at Estevan, and its members and support staff accepted an invitation to attend the weekly meeting of the local Rotary Club and outline the work that they were doing. Perhaps because it was March 17th, St. Patrick's Day, "Irish Charlie" Wilson was assigned the task of addressing the luncheon while Chief Justice Brown and R.W. Hugg, KC, looked on.

Charles explained the board's approach to ensuring that valuations were fair. The advice of the local member of the Soldier Settlement Board was frequently sought to assist in determining land values. When the value of all assets was fixed, the total debts were reduced to this level. Debts dealt with by the board so far had averaged a total of $10,000, from which an average of $4,000 had been cut.

The adjusted debt load was usually directed to be repaid over ten years according to an established formula: 3 percent of the new principal, plus interest, during each of the first three years; 5 percent and interest during each of the next three years; 7 percent plus interest in each of the next three years; and the remaining 55 percent of the principal, and interest, in the final year of the ten-year period. Interest was set according to the nature of the debt, with mortgages the highest at 6 percent. Banks and machine and lumber companies had to accept 4 percent.

Charles explained that the reason for spreading repayment over ten years in this manner was "the hope that farmers would be able to rehabilitate their homes and living conditions to an appreciable extent during the period of recovery." He then told the Rotarians of some of the serious situations that had come to the attention of the board.

> Mr. Wilson cited instances of farmers sleeping on gunny sacks stuffed with straw, eating in relays at tables where there were not enough dishes to go around, and dressing in rags that produced a sense of shame so strong that they would not appear when friends called at their homes.
>
> "The feeling of loss of ownership of his farm that has crept into so many farmers' hearts has to be eliminated and the rights of ownership properly respected."[8]

Back at Limerick, the office of Wilson Brothers, left in the hands of Kevin, was a concern to Charles. The office was still supervising the Victoria Trust mortgages, and Kevin was only twenty years old in 1937.

> *I knew the mortgage game called for the judgment of an older man. However, I underestimated my son. There was in the Bank of Montreal at Limerick at that time a very capable manager who later served the bank in Montreal, in Toronto, in Winnipeg, and in Vancouver and Calgary. Indeed, he retired as Alberta superintendent of the Bank of Montreal. My son formed an intimate relationship with him.*
>
> *The name of the bank manager was William C. Morley, and he was endowed with plenty of ability and plenty of personality. He alone in the village knew that my absence from the office could create a problem, and he formed the habit of visiting the office frequently in the afternoon and building up an intimate relationship with my son, Kevin. In two or three months, this grew to the point where Kevin, faced with a serious problem, took it up to the bank before deciding and talked it over with Mr. Morley. When I became aware of this, I commenced to sleep a little better in North Battleford and Prince Albert and the other points to which I was confined.*

But, when I managed to return to Limerick Saturday afternoon, I also formed a new habit. Instead of going down to the office and worrying about the individual problems, of which, of course, Kevin kept me fairly well advised in any case, I went to Mr. Morley's home and asked him the question "Do I have anything to worry about at the office?" He would almost invariably be able to assure me that I had nothing to worry about, that Kevin was really carrying the ball.

I doubt that any man incurs toward a friend a greater debt of gratitude than I incurred toward Mr. Morley.

The office of Wilson Brothers underwent more change. Lizzie Allonby, after twenty-two years of extremely capable service, decided that her age dictated retirement, and she moved to the more salubrious climate of Victoria. A new business school graduate from Regina, Olive Keating, was hired to fill the large shoes left by Miss Allonby.

As the *annus horribilis* that was 1937 drew to a close, the concerted effort to preserve the western farm population was never more in jeopardy. Two-thirds of Saskatchewan's rural people were on relief. Of the 302 rural municipalities, 290 were lined up seeking assistance from the provincial government, which itself was in trouble and in need of more support from Ottawa. Since the Depression began in 1929, Saskatchewan had spent much more on relief than its ordinary revenue, and most of the expenditure, of necessity, had come from the federal government.[9]

Ottawa also provided direct aid to the starving residents of the drought area in the autumn of 1937 by distributing 782 freight car loads of fruits, vegetables, fish, beans, and cheese.

Those who remained on the farms and endured the soul-curdling privation were not all men and women with an abundance of grim stubbornness. Many, likely most, were those who no longer had any place to go or the wherewithal to get there. They were stuck. To qualify for relief, one had to meet a residency requirement, and that alone discouraged movement.

There was not, as popularly thought, a migration from the rural areas to the towns and cities. In fact, most Saskatchewan urban centres lost population. As of the end of 1936, only three of the province's twelve cities and towns showed increased population, and of them only Prince Albert added an appreciable number. The other nine showed decreases.[10]

And the exodus out of Saskatchewan was not yet in earnest. But it had begun. Although the 1936 population of 931,547 was the highest yet, it should have been higher as some 65,000 people had already moved away.[11] By the end of the decade, it was estimated that at least 250,000 people had left.[12] Still, Saskatchewan retained the title of Canada's third most populous province, a distinction that it would not lose until after the Second World War.[13]

The thousands of abandoned farms and shuttered businesses testified to the outbound flight from the horror that was southern Saskatchewan. Single men could "ride the rods," and hardly a freight train could be seen during the summer months without its complement of drifters atop cars or, if lucky, in empty ones, the "hobos," "bums," and "tramps" of the era. Families travelled the roads, heading north or west, pitiful horse-drawn caravans, frequently with a tethered milk cow plodding behind.

As the winter of 1937–38 descended on Saskatchewan's misery, a normal snowfall brought a glimmer, not quite of hope, for that emotion was mostly stilled and no longer sprang from the western breast, but of a first tingle of curiosity, or wonder, that perhaps, just perhaps, the turning point had been reached. Surely no drought could last forever, and the snowfall on top of the fall rains just might signal the beginning of the end.

ENDNOTES

1. The phenomenon was repeated more than fifty years later, in 1988.
2. Timothy Egan, *The Worst Hard Times* (New York: Houghton Mifflin, 2006).
3. Ibid., 150–52.
4. Britnell, *The Wheat Economy*, 56.
5. Ibid., 78.
6. Ibid., 88.
7. G.E. Britnell, "The Saskatchewan Debt Adjustment Program," *Canadian Journal of Economics and Political Science* 3, 3 (1937): 375.
8. *Estevan Mercury*, March 18 1937.

9. Britnell, *The Wheat Economy*, 97–99.
10. LAC, Mackenzie King Papers, vol. 219, 188226, Saskatchewan Urban Municipalities Association brief to Prime Minister Mackenzie King, December 31, 1936.
11. Britnell, *The Wheat Economy*, 14.
12. Elizabeth Mooney, "Great Depression," *The Encyclopedia of Saskatchewan* (Regina: Canadian Plains Research Center, 2005), 414.
13. Saskatchewan's population did shrink after 1936, dropping to 831,728 in 1951 before slowly climbing back.

CHAPTER 19

1938

The spring of 1938 did bring a blessed end to the years of drought. The rain started early and continued through the growing season. April at Assiniboia saw 18.3 millimetres, or .72 of an inch. May brought 83.1 millimetres, or 3.72 inches. June was graced with 104.6 millimetres, or 4.11 inches. July saw 57.4 millimetres, or 2.25 inches. Through August, September, and October came another 99.1 millimetres, or 3.89 inches.

The crops luxuriated. The pastures greened. Hope returned.

Then came the cruel reminder that drought is but one of the many plagues that beset those who would wrest a living from the soil. First came another tsunami of stem rust. Next hail storms so vicious and extensive that in just one week the Saskatchewan Wheat Pool dropped its production estimate from 218 million bushels to 190 million bushels.[1] Then again came the grasshoppers in several airborne invasions from the United States, and this time Regina and its residents bore the brunt of these assaults. On July 22nd, the insects descended on the downtown area in clouds so dense that pedestrians scattered, into doorways, anywhere to escape the pests.[2] On August 1st, residents watched with relief as the third great flight of 'hoppers in ten days passed over the city heading north.[3] But on August 10th, Regina was chosen again, and according to the *Leader-Post* grasshoppers "took over the city."[4] It was described by James Gray as "[t]he worst grasshopper blizzard within the memory of man."[5]

The splendid crops could not withstand all of these attacks. The average wheat yield across Saskatchewan fell to a disappointing 9.6 bushels. Total production was only 132 million bushels, the fourth best since 1929, but a long way from the early estimate of 218 million bushels. In September, Minister of Agriculture J.G. Taggart claimed that grasshoppers and rust had combined to rob Saskatchewan farmers of at least 50 million bushels of wheat.

Taggart also announced a program to exchange, bushel for bushel, current stocks of seed wheat for the new rust-resistant Thatcher strain. In the spring of 1938, Premier Patterson called an election and chose the date well—June 8th. Then the province was full of promise, and the rust, hail, and grasshoppers had not yet appeared. The Liberals won thirty-eight of the fifty-two seats. The Conservatives, under their new leader, John G. Diefenbaker, were able to field only twenty-four candidates and came up with just 12 percent of the vote and no seats. The Cooperative Commonwealth Federation (CCF) won ten seats and remained the official opposition. A powerful onslaught by Premier Aberhart of Alberta and his Social Credit introduced a lot of furor to the campaign but produced only two seats. Two Unity candidates were also elected.

The duplication of agencies and efforts that Attorney General T.C. Davis had both predicted and deplored in his 1934 correspondence with M.A. MacPherson in Ottawa had come to pass. In addition to the official receivers and the Board of Review operating under the FCAA, the provincial Debt Adjustment Board was actively intervening between debtors and creditors. Although without the power to order debt reduction, as the Board of Review could, the Debt Adjustment Board had the power of moratorium and could prevent mortgage companies from realizing on their security. Board member George Edwards explained the process in his memoirs, written in 1969:

> A special Debt Adjustment Act was passed which empowered the Board to prevent any action from taking place without the consent of the Board. If a mortgage company wished to take

action against a farmer to foreclose a mortgage or cancel an agreement for sale or take action against a home owner who was behind with payments on the home, they were obligated to give the debtor notice of intention to take such action and our Board received a copy of this notice of intention. We immediately notified the debtor that if he wished protection against the action to contact as soon as possible the sheriff for the judicial district in which he lived and give him a full statement of his affairs and we instructed the sheriffs to see to it that no one who was doing what could reasonably be expected of him was to be dispossessed of his home or farm. The responsibility for deciding whether the action should be allowed or not allowed rested with the Debt Adjustment Board. The purpose of the Act and the instructions we received from the Government were to protect deserving debtors no matter how much they owed if they, on their part, were doing their best.

I personally had the responsibility of deciding practically all farm cases, although I could consult other members of the Board. However, in actual practise I rarely did this. This was a very onerous task as the financial distress was almost universal as the crop failures or partial failures, as well as the world wide depression, lasted all through the thirties and well into the forties. My work often entailed personally visiting the farmers and thus getting a first hand picture of the whole situation in which the farmer found himself.

By the time the work of the Board was finished in 1943 we had a total of one hundred and twenty thousand files.[6]

Mr. Edwards is entitled to a very large place in the history of this province in my opinion. When the Gardiner government came to power, he was naturally selected as a member of the provincial Debt Adjustment Board and went to work in that office with the fierce enthusiasm which must frequently have been a source of embarrassment to his colleagues. I doubt that he would have remained there very long had it not been for my personal influence with him. I remember telling him one day, "George, if you remain here for two years and succeed in that time in accomplishing

*as much as you hope to accomplish in the first two weeks, you will really
have done a man-sized job."*

Additional agencies were created to tackle the farm debt monster
and its bureaucratic tentacles. Even with the most agreeable creditor,
reducing mortgages or even unsecured liabilities required much more
than a handshake. Paperwork was necessary. The Voluntary Adjustment of Debts office was established in the Municipal Department
under the direction of a senior lawyer, A.S. Sibbald, KC.

Acknowledging that the machinery initially created by the FCAA
was moving too slowly to cope with the volume of work before the
Boards of Review, the federal government amended the legislation to
permit the establishment of additional boards. In 1938, second boards
were appointed in Alberta and Saskatchewan, the latter under the
chairmanship of Court of Appeal justice William Martin, a former
premier.

Even with this additional effort, the crusade to bring the burgeoning
farm debt under control was losing ground. At the end of 1938, taxes,
interest, and relief had taken the total farm debt back to $525 million,
the 1936 level. It was a matter of considerable concern.

When Prime Minister R.B. Bennett introduced the FCAA to the
House of Commons on June 4, 1934, he explained that Canada was
facing a total farm debt of $726,026,500. Saskatchewan's share was
$175,770,300.[7] His figures were based on the 1931 census and, he ad-
mitted, were not precise. Nonetheless, four years later the farm debt
in Saskatchewan was almost exactly three times what it was thought
to have been when the FCAA came into being.

Minister of Agriculture Taggart, noted for his plain speaking,
outdid himself before the annual conference of the Canadian Institute
of Economics and Politics at Lake Couchiching, Ontario, on August
17th. "It doesn't matter a bit" who governs Saskatchewan, he said. "No
government could do much harm or good. No government could bring
us more trouble than we have now. We couldn't lose much if we lost
all, and we couldn't get further into debt because no one will loan us
any money."[8]

Of Saskatchewan's total agricultural debt of $500 million, Taggart said that only $180 million was owing on mortgages. "The farmers owe more to other farmers and individuals than to corporations. They owe practically nothing to banks."[9]

Relief was a continuing problem as 1938 saw the expenditure of another $50 million to support the rural population. Alarmed, the Saskatchewan government adopted a policy "to encourage farmers to retain all living and operating expenses (including seed, feed and fodder, etc.) for the coming year out of the low returns of the 1938 crop, and to pay one year's municipal taxes, before attempting to make any payments on debt account."[10] The Debt Adjustment Board applied the policy in its interventions between creditors and debtors.

In an attempt to provide employment to the thousands of single men cut adrift by drought and depression, the governments in Ottawa and Regina instituted the share-costed Farm Improvement and Employment Plan. The scheme provided a subsidy for the winter employment and boarding of men on operating farms, usually looking after cattle. The assistance was minimal, ten dollars a month split between employee and employer, but between October 1, 1936, and February 15, 1938, 58,277 applications were received for placement.[11]

The return of the rains continued into 1939. At Assiniboia again, April saw 20.3 millimetres, or .8 of an inch; May brought 67.6 millimetres, or 2.6 inches; June a whopping 163.6 millimetres, or 6.4 inches; July 42.2 millimetres, or 1.7 inches; with another 39 millimetres, or 1.5 inches, during the fall. With the other plagues of rust, hail, and 'hoppers at moderate or low levels, the wheat crop responded with an average yield of 17.1 bushels and production of 250 million bushels, the largest since 1928. Unfortunately the farm price dropped almost in half, from $1.03 to fifty-eight cents.[12]

In June 1939, Ottawa enacted the Prairie Farm Assistance Act (PFAA), which provided a limited form of crop insurance. Farmers were charged 1 percent on their grain deliveries, which entitled them to collect $2.50 an acre on half of the cultivated acreage when yields dropped below five bushels per acre. The PFAA ,with various modifications, continued as a vital element of prairie grain farming until it was overtaken by crop insurance some twenty years later. Even so, PFAA was not repealed until 1985.

On June 29, 1939, T.C. Davis was appointed a justice of the Court of Appeal, and he resigned as attorney general and MLA for Prince Albert.

It happened that we were both up at Waskesiu, and I dutifully called on him to congratulate him on his appointment, and then I chided him for having retired to a monastery and taking himself out of the active life of the province which he had so much adorned. His reply was a characteristic one. He said, "Charlie, it's nice to know where you are going to eat in your old age."

Some months later he became chairman of the Board of Review,[13] of which I was a commissioner. I met him at the Assiniboia Club, and again I dutifully congratulated him upon the appointment. That done, I told him, "The truth is I do not really approve of your appointment to this board."

He asked me, "In the name of God, why, Charlie?"

I told him, "I will tell you very frankly. I know you to be a man of great heart and large compassion, but you were born in the purple, and you have never been broke in your life. With all your attainments, I gravely doubt that you will appreciate how these people out on the back road allowances are circumstanced and are living."

His reply to me was "Oh! That's what you're thinking. I can tell you something, Charlie Wilson. You and I are going to get along splendidly on this board."

He was right. I had the most pleasant relationship with him on the board, and I found that you could disagree violently with him without creating any difficult situation and that he would listen with the most meticulous care to your side of the argument and was not a little slanted in favour of the unhappy debtors with whom we were dealing.

We were only able to hold Davis for a short period. The work was not sufficiently intimately connected with the war to be satisfying to him, and one morning he turned up as deputy minister of war services at Ottawa.

The footprints which Tommy Davis left in this Province of Saskatchewan were deep enough that I do not feel called upon to try to enlarge them. They should endure for a long time, for his was, indeed, a distinguished career in this province and outside of it.

In the place of Davis, there came to us Justice Percy M. Anderson, and a better man could not have followed Davis. Judge Anderson was a man of compassion and understanding in the highest degree, and my relationship with him for the balance of my term on the board was a very warm and cordial one indeed. He continued to be our chairman until the board was dissolved in 1943.

One evening I boarded CPR No. 3 at Toronto to return to Saskatchewan after the Christmas holidays. To my delight, I found my old friend, Mindy Loptson,[14] the MLA for Saltcoats, on the train. The next evening, in the dining car, when we were through, Mindy leaned over and picked up my dinner check. I immediately told him to put it right back where he found it since I was travelling on an expense account anyway. But Mindy's reply was "You made me a saving of $3,000, and I want the pleasure of buying you a dinner in return."

I said, "Mindy, that statement is not true. You never appeared before our board, so we had no opportunity of saving you $3,000."

"That's so," he said, "but my neighbour across the road from me appeared on similar land having his mortgage with the same company. When he received your judgment, I borrowed it from him and went to Winnipeg and had an interview with the company. I placed your judgment on the manager's desk and then asked him, 'What is the sense of causing both of us the trouble of going all through that procedure when it is probable that we know the results in advance?'

"The manager agreed with me and wrote off the same amount from my mortgage that you had written off my neighbour's, $3,000, and I insist on buying this dinner for you."

There must have been countless cases of voluntary adjustments made under the same circumstances as Mindy Loptson's. At any rate, the work commenced to lighten a little, and we were sometimes in hope that we might finally see some daylight.

ENDNOTES

1. James H. Gray, *The Winter Years* (Toronto: MacMillan of Canada, 1966), 115.
2. *StarPhoenix*, July 22, 1938.
3. *Leader-Post*, August 1, 1938.
4. Ibid., August 10, 1938.
5. Gray, *The Winter Years*, 115.
6. Edwards, "Memoirs of George F. Edwards," SAB A-3,29.
7. *Debates*, June 4, 1934, 3649.
8. *Leader-Post*, August 17, 1938.
9. Ibid.
10. Britnell, *The Wheat Economy*, 88.
11. *Saskatchewan Legislative Journals*, February 16, 1938.
12. C.F. Wilson, *A Century of Canadian Grain* (Saskatoon: Western Producer Prairie Books, 1978), 246.
13. Justice Martin resigned from the Board of Review upon being appointed custodian of enemy alien property in Ottawa. The Davis board operated in parallel with the board chaired by Chief Justice Brown.
14. Asmundur Loptson represented the Saltcoats constituency from 1929 to 1960 (and Pheasant Hills from 1934 to 1938). In 1953–54, he served as house leader of the Liberal opposition in the legislature.

CHAPTER 20

1939: RAIN RETURNS

The armies of Adolph Hitler invaded Poland on September 1, 1939. Great Britain and France declared war on Germany two days later. Canada followed one week later, on September 10th. The nation was at war. Military recruiting offices opened across the country. Federal and provincial programs subsidizing the employment of single young men would no longer be needed.

The Second World War had an early impact on the western Canadian grain trade. Traditional markets in continental Europe were cut off, leaving England as the sole importer of wheat. Later the United States became a major customer, though it too was already a large wheat exporter.

Western farmers, who for ten years had struggled against drought, rust, and grasshoppers to produce wheat, now found the markets for that wheat disrupted by events in far-off Europe. To cope with the consequent grain carry-over, Ottawa instituted a quota system of grain delivery from the farm based on acreage. And the initial price of wheat was pegged at seventy cents a bushel, Fort William.

Prime Minister Mackenzie King called a federal election for March 26, 1940. With the war in the background, most Canadians rallied around the Liberal government, increasing its seat total by six to 179. The Conservatives lost three seats, dropping to thirty-six, while the

CCF scored an increase of one, to a total of eight. In Saskatchewan, the Liberals lost four seats, ending up with twelve. The CCF added three for a total of five. The Conservatives won two ridings, one of them, Lake Centre, by a lawyer named John Diefenbaker.

The rains continued to grace Saskatchewan, and wheat production responded. In 1940, Assiniboia received 39.1 millimetres, or 1.5 inches, of precipitation in April; 40.1 millimetres, or 1.5 inches, in May; 94.5 millimetres, or 3.7 inches, in June; and another 76.7 millimetres, or 3.0 inches, in July. The provincial wheat yield was 17.1 bushels per acre, slightly below that of 1939, but an increased acreage produced 266,700,000 bushels, the largest since 1928.

It was a good year to venture into farming, and in 1940 Victoria Trust and Savings Company did just that. The company had a large number of loans in the Killdeer region, south of Wood Mountain, an area that had suffered mightily during the drought, and many of the loans had gone bad.

It was many years behind Limerick in its development, and the people were not so tightly attached to the land, a factor which we all overlooked. When the Thirties descended upon us, a great many of those settlers threw up their hands and moved to what they thought were greener pastures, with the result that we had a lot of land, something in excess of 4,000 acres, on our hands in that area.

I was afraid that, if we foreclosed our mortgages and took title to the land, the municipal authorities, destitute of money, would be likely to sue us for the taxes, and so we allowed the land to sit as it was, with the mortgage on it, and only in one or two cases did the company become owner of it.

Even so, by 1940 the taxing authorities were becoming pretty restive, and it was clear that the situation would not last for any great length of time. There were no buyers available, and even very few tenants, and the problem was an acute one. I just did not know where to turn for a solution.

One day Kevin came to me with a solution all tight, the solution being that we should go into possession under our mortgages and farm the land on behalf of the company. It proved very difficult for Kevin to sell me that idea. I urged upon him that if the storm was going to last much longer it could easily prove to be better business to forget about our mortgages in that area and let the taxing authorities take title if they wanted to. I mentioned one cardinal feature. I said, "You are proposing to start farming 4,000 acres. That's a big undertaking, even in Saskatchewan. Your success will depend upon getting a competent farm manager, and I do not see where you can get him. If he is that competent, he is no doubt self-employed."

Kevin's immediate reply was "Dad, I have the man. Henry Dion. Henry has the ability and the know-how and the energy. All his life he has lacked capital. We propose to furnish the capital, and I am certain that everything will work out."

Finally he got a grudging assent from me that I would submit the scheme to head office. We fully expected it would be declined. It was a rather fantastic adventure for the little quiet Victoria Trust and Savings Company. When I submitted it, I was astounded by the alacrity with which it was accepted by the Board of Directors. We put in our first crop in 1940—1,600 acres.

Of course, the lands had been terribly neglected during their abandonment, and there was no summerfallow and no land that could really be declared fit for cropping. However, with brand-new machinery, Henry Dion went over that land in spite of a three-foot crop of Russian thistle on top of it, rarely employing fire to clear the land because fire was apt to lead to drifting soil. But with the new machinery, he was able to get 1,600 acres of crop in, and when it was done there was no more sign of black land than before he started. The land was covered with a six-inch thatch of Russian thistle and other weeds, through which the young wheat had to force itself.

On the 1st of July 1940, I went down to see what was happening, and I was so disheartened with the first quarter section I inspected that I let the rest go. The young wheat was slowly forcing itself through the thatch, but the date was July 1st, and I did not think that it had a reasonable chance of reaching maturity.

I did not see it again until harvest time, but before harvest arrived we were running ourselves ragged trying to find buildings to house the expected crop. I did manage to induce my friends in the Wheat Pool, who were building 25,000-bushel annexes to most of their elevators in

Saskatchewan—for 1940 was a banner year—to build us an annex adjacent to their elevator in Killdeer and to connect it with their elevator by a long spout from the cupola of the elevator to the centre of the annex. This took most of the problem off our hands, and we realized a crop of approximately 25,000 bushels from our first venture.

The two Saskatchewan Boards of Review continued to grind against the province's farm debt, making slow but finally discernible progress. Since 1936, they had used 6 percent as the standard rate of interest applied to the mortgage loans that they dealt with, though their counterparts in Alberta and Manitoba generally adhered to a 5 percent rate. The higher Saskatchewan rate was justified because its boards included in all of their proposals, or orders, a crop failure clause to protect the farmer in case of his inability to meet the terms of the proposal. In 1940, the courts struck down the boards' authority to include the crop failure clause. Charles was then working with Chief Justice Brown's board.

Our new proposals were accumulating interest which could not be paid. One afternoon I pointed this fact out to the board and then made a proposal that the interest rate should be reduced still further to 5 percent as somewhat of a concession to the situation. The chief justice looked me hard in the eye for a moment or two and then said, "Charlie, you are asking for a 5 percent rate of interest. You and who else?"

This was an excellent question, socially tied in with politics, in which field the chief justice himself had no mean experience. I got the import of the question quickly and replied, "I withdraw that resolution for the moment."

I then proceeded to the Wheat Pool building and told the board there what had come up. I reminded them that they had promised me cooperation, and this was the first occasion on which I needed it. I also reminded them that I could not go out in the country weekends and hold meetings with the farmers and get resolutions from them and then come back to Regina and adjudicate on the resolutions which I had sponsored. My office was supposed to be a semi-judicial one and was surrounded by pretty strict rules of conduct. I pointed out to the Wheat Pool that they had an organization all over the province and could easily hold meetings and obtain resolutions. They promised me that they would do so without delay and did.

A couple of weeks later they presented us with a very well-prepared brief pointing out the surrounding circumstances and asking for a rate of interest of 4 percent. I was now able to put my resolution to my board, reminding the chief justice of the question he had asked me and telling him, "I now present the matter to you on behalf of myself and the organized farmers of this province."

I got the consent of the judge to my 5 percent rate.

About the time we started to work, the Supreme Court of Canada made a ruling that we could not ignore security or take security away from a creditor. This obliged us to leave with the loan company the full amount of the equity which we considered it had in the land. In most cases, there was a tremendous amount of debt beyond that equity, with little or no security, possibly a lien on some well-worn machinery.

We were conscious of a volume of debt which we really hated to have to reduce in the hands of the general storekeepers of the province who had continued to furnish groceries to the farmer clients long after they should have discontinued. We were quite well aware of this fact at the time, but, if we were to open a door for the farmer, we had no alternative except to effect a drastic reduction in the amount of the unsecured debts.

I am taking this opportunity of reminding the Province of Saskatchewan of a great debt which they certainly owed at that time to their local merchants. While I had great sympathy for these local merchants, I did not see anything that I could do except to plow ahead on the established lines and treat them as ordinary unsecured creditors.

Back at Limerick, Kevin began to assume more and more responsibility for the Wilson Brothers office and the affairs of Victoria Trust and Savings Company.

I decided that he might as well have a free hand all across the board, and in that I proved to be quite justified. The office never pulled a blooper so long as it was under Kevin's direction.

The outstanding success which Kevin achieved in the Killdeer area suggested to us a policy which led us into a wider field. Under the difficult conditions prevailing until 1942, the company had crop leases with every borrower entitling it to one-third of the crop. Many of these crops had a tendency to disappear between the fall and spring, and Kevin got another idea. He organized a group of young farmers at Limerick who had trucks, and late that fall they called at all these farms, picked up the company's share of the crop which the farmer was unable to sell, and transported it by the CPR to Kisbey. We had to pay local freight on the wheat from the local points to Kisbey because the Crow's Nest Rate only applied to terminal elevators, and, of course, that was done. At that point, the company had acquired a big stock farm with immense barns on it and was in a position to store 100,000 bushels of wheat. Those buildings were renovated and wired securely, and we ended up that fall with 100,000 bushels of wheat in store at Kisbey. We still believed at Limerick that good sound wheat in a good sound bin was almost the equivalent of money in the bank, and our confidence proved to be fully justified.

A couple of years later conditions had altered to the point where we were able to ship the wheat to Fort William, but that raised a difficulty in respect to freight rates. The CPR rule at that time was that you could break the trip with a carload of wheat, but the trip had to be resumed within six months in order to get the benefit of the through rate from the point of origin to Fort William. The difference was a very, very substantial one, and we were going to be crowded to the limit—in fact, we were going to find it impossible—to get the wheat out in six months, in view of the fact that it had already reposed at Kisbey for a period longer than that.

When the wheat was ready to move, Kevin brought the problem to me and said, "It's up to you to find a solution that will induce the CPR to lift that rule and give us the benefit of the through rate even though the period exceeds six months."

I thought for a moment and told him, "I don't think I'll have a bit of difficulty with that. My good friend Colonel James Cross is chairman of the Board of Transport Commissioners at Ottawa, and the relationship between us is close enough that I think he will be very happy to make that arrangement for me."

I immediately wrote to Colonel Cross and got a reply from him on the stationery of the Transport Commission regretting that freight rates were quite outside their jurisdiction, and there was nothing at all he could do to be of assistance.

This threw me back on my own resources, and I went to Regina in a few days and paid a call on the district freight agent. When I secured an interview with him in his office, I was surprised and delighted to find that he was a gentleman with whom I used to play a lot around the Assiniboia Club, William Anderson. I had known he worked for the CPR, but I did not know what his job was.

I expressed my delight at finding him in that chair and told him, "Bill, I have a real problem, and I really want your help." I then outlined the problem to him. Mr. Anderson looked at me with a rather quizzical smile on his face and said, "Charlie, isn't that problem all adjusted?"

I said, "You would make me very happy if you were able to tell me that, but not to my knowledge."

"Isn't that Colonel Cross at Ottawa your agent?" he asked.

I said in astonishment, "He is my warm personal friend, but neither I nor the Victoria Company would presume to call the chairman of the Transport Commission our agent. In any event, Colonel Cross wrote me that in his department he had no jurisdiction and could do nothing for me."

He said, "Well, what he did was he hopped on the train and went down to Montreal and fixed it the way you want it. I have here two copies of two letters which I am not authorized to release to the press until the day after tomorrow, but I am going to permit you to read them now."

He did so, and the problem was completely solved, and I had no further worries.

That problem was quickly replaced by a new one. The CPR had a rule, and I expect they have it still, that if a car is unloaded en route, if it is going to receive the through rate on reloading, it must be reloaded to exactly the same weight as when it was unloaded. This was again a real problem, and I could not figure out a way of getting around it. But I went to the Wheat Pool building in Regina and interviewed my warm personal friend, William Lawless, who was in charge of the Pool's general country operation.

I showed Mr. Lawless the correspondence, and he took the letters from me and told me, "Charlie, that's altogether too sticky a matter for you. I want you to leave those letters with me. I'll look after this problem for you and make you a return when the wheat has all been loaded."

I was only too happy to follow Mr. Lawless' suggestion, and about a month later I received from him a return about the size of a large blotting pad, showing every car as it was unloaded, every car that was reloaded, with the weights and the freight. Mr. Lawless had even paid the freight to Fort William, so that all I had to do was sign a draft on the company for the amount of money that was owing him for the freight.

The wheat was actually loaded at a little point called Armilla, midway between Kisbey and Arcola, where the Pool—most fortunately for us—had an elevator.

As 1941 opened, the Second World War was going badly. France had fallen, and the German Army had overrun Europe. Although the Luftwaffe had been staved off in the Battle of Britain, England was very much a besieged nation. The United States remained neutral, a posture that it would maintain until December 7, 1941, when Japan attacked Pearl Harbor. Even then, America tried to avoid the war in Europe; it declared war against Japan only, until December 11th, when, strangely, Hitler declared war against the United States.

They were dark days for the western democracies as they frantically geared up their military strength. Residents of southern Saskatchewan became accustomed to the snarl of Harvards and Cornells from the several bases of the British Commonwealth Air Training Plan.

In February 1941, Kevin enlisted in the Royal Canadian Air Force. A new manager was required for Wilson Brothers and the Saskatchewan operations of Victoria Trust and Savings Company.

I approached a friend in Regina, Mr. William H. Gundry, who had spent a lifetime in the service of the Trust and Guaranty Company. I asked him to come to Limerick and take over the office so soon to be vacated by Kevin on account of his enlistment. Although the salary was not a large one, Mr. Gundry accepted with alacrity, telling me that he was very proud to get an opportunity to replace a man in uniform.

He came to Limerick and was a man of extremely sound judgment, not possessed of much imagination, but insisting on dealing with the underlying facts in every situation. He had the disqualification that his right leg was a wooden one, which he did not permit to interfere very greatly with his locomotion at that. There was nothing wrong with his integrity and his determination. He stayed for quite a few years as manager of the Limerick office and proved to be a quite invaluable asset in my life.

CHAPTER 21

AN ADVENTURE IN WHEAT

In 1941, the rains faltered somewhat, and wheat yields fell well below the two good years of 1939 and 1940, down to an average of twelve bushels per acre. In response to federal government incentives, 3.5 million acres were taken out of production in 1941, so that total Saskatchewan production of wheat fell to 147 million bushels. With an elevator price of roughly fifty cents a bushel, economic distress re-entered farm communities. As usual, southern Saskatchewan suffered the greatest loss.

Rural discontent spawned another trek to Ottawa. Organized by the Saskatchewan Wheat Pool, two special trains carried 400 concerned farmers into Ottawa on February 1, 1942. The King government received them with sincere attention. The plan was to invite the delegation onto the floor of the House of Commons, as had been done in 1910, but because Parliament was in session the show was moved to the convention facilities in the Chateau Laurier Hotel. There, on February 2nd, Prime Minister King and nine of his cabinet ministers listened to the western spokesmen, who were supported by a petition carrying 185,000 signatures requesting that the wheat price be a dollar a bushel.

A high degree of mutual respect and consideration was maintained between the delegates and the government representatives. That evening all twenty-one Saskatchewan MPs, which included Prime Minister King, entertained the delegation at dinner in the parliamentary restaurant, an invitation that was reciprocated the following evening when the delegates treated the MPs in the Chateau Laurier dining room.[1]

In the end, the wheat price was not achieved, but the government did direct the Wheat Board to increase the initial payment from seventy cents to ninety cents.

The 1942 crop removed practically all the discontent. Rain blessed the land. Assiniboia recorded 62.0 millimetres of precipitation in April, or 2.4 inches; in May, 49.8 millimetres, or 2.0 inches; in June, 123.4 millimetres, or 4.8 inches; in July, 60.2 millimetres, or 2.4 inches; in August, 113.0 millimetres, or 4.4 inches. The land responded with an average yield of 24.7 bushels per acre and a total crop of 305,000,000 bushels, again the largest since 1928.

Saskatchewan was still a long way from prosperity, but a taste of liquidity was returning.

The work of the Boards of Review across Canada began to wind down. Ottawa amended the FCAA, closing off further applications under the act in all but the four western provinces after December 31, 1938, and in Manitoba and British Columbia after June 30, 1939. Only in Alberta and Saskatchewan could insolvent farmers still make proposals, and even in those provinces the volume began to decrease. By the spring of 1942, the Saskatchewan Boards of Review still had 409 cases in front of them, while the Alberta boards had 282.[2]

The work commenced to lighten a little, and we were sometimes in the hope that we might finally see daylight. The volume of applications was slowly declining, until by 1943 the reduction was quite noticeable.

There were bright spots. I remember a farmer's wife writing us to inform us that it was their household custom to have evening prayer and that from now on the names of our board would be included in those prayers and another farmer who wrote us that he and his wife had decided to entitle our proposal their new charter of liberty.

Incidents like these encouraged me to believe that we were breaking through the fog which rested so heavily on our province and that a little daylight was commencing to appear at the top. It was becoming possible to find tenants again for vacant lands because farmers were regaining the hope that a little profit could be made out of farming, a way of life to which they had become attached and which meant almost everything to them.

The powers which we had been exercising as a board were conferred upon the judges of the District Court, who, following our guidelines which we had laid down, made very good success of cleaning up the balance of the problem, and we were in 1943 released from our duties.

The Farmers' Creditors Arrangement Act of 1943 supplanted the 1934 act and its amendments. The Boards of Review ceased to exist, and their work was taken over by the District Courts. As prosperity returned to the west, diminishing use was made of the procedures of the FCAA. The legislation was repealed in 1988.

Even today the total volume of the debts which we removed from the farmers of this province would be a great shock to a great many people. At some of the courthouses where we held our sessions, our average rate of reduction was maintained at 48 percent, and that's in a province where there was no precedent for any reduction of debt, much less the extremely drastic reductions which we found ourselves forced to grant to the farmers if they were to have a hope of emerging with clear title to the land at any period in their lifetime. I am aware that a large number of them achieved clear title quite a few years back, and I am gratified by that fact.

I think it is an absolutely fair statement to make that I carried more than my share of the load and that my seven years' work on the board constitutes by long odds the greatest service which I had the privilege of rendering to my province.

There were 8,235 applications to the Saskatchewan Boards of Review to April 1943. The reductions totalled $35,850,960 for an average of 44.33 percent. In 2012 dollars, that would represent more than half a billion dollars of debt wiped away (see Table 21.1).

THE FARMERS' CREDITORS ARRANGEMENT ACT, 1934.
DOMINION OF CANADA—PROVINCE OF SASKATCHEWAN
Return
(BY JUDICIAL DISTRICTS)

Showing Gross Debts of Applicants For Whom Proposals Formulated By The Boards of Review and Amount of Reductions Made Available Under The Boards' Proposals For Periods to 30th April, 1943.

DISTINGUISHING NO. AND NAME OF JUDICIAL DISTRICT	NUMBER OF PROPOSALS FORMULATED	GROSS DEBTS REVIEWED BY BOARDS	REDUCTIONS MADE AVAILABLE	PERCENTAGE OF REDUCTIONS
1. Arcola	198	$1,872,864	$912,126	48.70%
2. Assiniboia	644	7,576,419	3,353,128	44.26%
3. Battleford	851	6,570,881	2,859,354	43.51%
4. Estevan	168	1,635,086	709,495	43.40%
4a. Gravelbourg	419	6,595,508	3,283,092	49.79%
5. Humboldt	654	4,239,560	1,671,259	39.42%
6. Kerrobert	284	3,510,235	1,600,483	45.59%
7. Kindersley	304	3,726,911	1,788,612	47.99%
8. Melfort	341	2,327,887	865,910	37.19%
9. Melville	282	1,898,654	810,627	42.70%
10. Moose Jaw	252	4,140,512	1,954,921	47.21%
11. Moosomin	312	2,519,755	1,105,908	48.88%
12. Prince Albert	748	4,889,135	1,882,066	38.5%
13. Regina	444	6,471,540	2,825,425	43.66%
14. Saskatoon	465	5,265,398	2,513,579	47.73%
15. Shaunavon	302	3,370,340	1,415,514	42.00%
16. Swift Current	325	3,938,931	1,775,632	45.08%
17. Weyburn	180	2,693,072	1,227,768	45.59%
18. Wilkie	290	3,020,302	1,456,390	48.22%
18a. Wynyard	226	1,534,154	655,099	42.07%
19. Yorkton	546	3,075,111	1,184,572	38.52%
Total	8,235	$80,872,255	$35,850,960	

Average reduction 44.33%

Table 21.1

Some years ago, at the bus depot in Assiniboia, a farmer approached me with outstretched hand, a farmer whom I did not recognize. He said, "You will not know me, but I remember you. You put me on my financial feet."

I said to him, "Sir, there are still two opinions about the quality of the work which we did on the Board of Review. One school maintains that we cancelled far more than necessary, and a small group still contends that we did not go far enough. What is your idea of it?"

He said, "Mr. Wilson, you went to the right point. You could have left the farmers with a debt which would have taken a whole lifetime to discharge, instead of which you cut it to the point where a careful man could work out in five or six years, and most of us have succeeded in doing that. It would be my opinion that you stopped at the right spot. I want to point this out to you. As a result of your work, we are now in a position to buy new tractors from Ontario, new combines, new equipment, and contribute to the recovery of business all across Canada. I am sure that your judgments were correct."

In 1939, Moira Wilson graduated from the University of Saskatchewan with a bachelor of arts degree, majoring in English. Along the way, she had picked up business training with typing and shorthand. Thomas Donnelly, the local MP and a family friend, enlisted Moira as a secretary in his Ottawa office. Not long after, she was recruited as a secretary to the high commissioner at the Canadian Legation in Washington (in January 1943, it became the Canadian Embassy), where she remained during the war years.

When Kevin Wilson joined the RCAF in early 1941, he had been selected for pilot training but soon discovered that his eyes lacked the depth perception essential to successful landings, and he transferred to the Air Observer, later Navigator, course. Upon graduation, he was commissioned as a pilot officer. In October 1941, he married Olive Keating, secretary in the Wilson Brothers office.

In April 1942, Kevin was shipped overseas, where he served as a navigator on a squadron of Wellington two-engine bombers and participated in fourteen raids over occupied Europe. Then his crew was transferred to #420 Snowy Owl Squadron and moved to Tunisia for operations in support of the planned invasion of Sicily. On June 27, 1943, his plane, with an experienced, all-commissioned crew, failed to return from its first mission against targets in San Giovanni.

Kevin Wilson and Olive Keating with Dick the Crow in front of Wilson Brothers, 1940. Photo in possession of the author.

Florence was hosting several ladies at afternoon tea when she spotted Clyde Murray, the CPR station agent and telegrapher, coming up the walk. Instantly sensing his errand, she met him at the door before he rang the bell, wordlessly took the telegram from his hand, tucked it into her apron pocket, and continued with her hostess duties until her guests left. Only then did she open the telegram and impart the sad news to Olive, her daughter-in-law, and by telephone to Charles, who was in Regina. "Missing in action," said the telegram.

Charles refused to see finality in the news. He wrote to Moira in Washington.

We are all behaving good in the matter of Kevin, and I hope you are doing the same. For my own part, so strong is the conviction in my heart and bones that he is safe that I am just incapable of fear and worry and have no difficulty in concentrating on my work. And I know you won't attribute that to indifference, for you know Kevin and I were very close together. It will no doubt be some time before we get definite news. The Italians have other things to work at just now than making out lists of names. In the meantime, a certain amount of anxiety is unavoidable, and that just has to be endured. There are a lot of other homes in similar plight.

The Wilson family clung to a vestige of hope even after the second telegram arrived. "Presumed dead," it said. But there was nothing more, not ever, not even in the war records of the Italian and the German forces who might have claimed credit for an Allied plane shot down. Just silence.

Later in that summer of 1943, Charles returned to Limerick, his work with the Board of Review at an end.

When I came home from the Board of Review in 1943, I went down to the office, and Mr. Gundry immediately said, "I expect you want this chair of mine back."

I said, "No, Mr. Gundry, and I won't for some time to come. I want you to put me on the payroll for a pittance. I am going out in the country in order to get familiar over again with our borrowers and with the conditions in the country."

I did that, and one afternoon I went into the office and asked him, "Mr. Gundry, what are these lands at Limerick good for?"

He replied, "They are good for growing wheat—nothing else that I know of."

"If you are unable to sell the wheat, what are the lands good for then?"

"If you are unable to sell the produce off land, then the land has no value," he replied. "Come clean. What idea have you got in your head now?"

"This one," I said. "I find out in the country that our farmers have all kinds of wheat. They took off a splendid crop in 1942, more than they could put in their granaries. Much of it is piled outdoors yet, and they are threatened with another good one in 1943. They would be delighted to give Victoria scads of wheat in the hope of getting it applied on their debt and in the meantime in the hope of getting a roof over its head.

"Do you think it would be sound if we rented, or obtained by any means, storage and took this wheat as collateral to our mortgages, banking on being able to find a market for it in the coming years?"

He did not hesitate. "That policy is just as sound as a brass bell. I am convinced that this wheat has value, and that there will be a market for it, and that before too many years have elapsed."

"*Very well. Since the suggestion has your approval, I'm going to submit it to my Board of Directors at head office and see what they think of it.*"

I was fortunate in having a Board of Directors who had confidence in the province, and they accepted the policy without any hesitation and told me to go ahead and get my hands on all the wheat I could, a very simple job, as it proved to be.

Immediately we rented every vacant building that we could find in the south country. There were two big garages in Limerick, unoccupied, each capable of holding 22,000 bushels. We secured them and a third one in the Village of Meyronne capable of holding an equal amount. We had our own annex at Killdeer, capable of holding 25,000 bushels, and our big barns at Kisbey were empty and capable of holding 100,000 bushels.

Through our good friends at the Wheat Pool, we were enabled to purchase twelve big Lockstave bins in Vancouver, each with a capacity of 10,000 bushels, the last twelve that were available, as it happened. We had them shipped to the various points that we had decided upon and erected, and they served our purpose admirably.

What this means is that we went into the elevator business, and nobody around the office had any experience in that game. We got books of storage tickets printed and were fortunate to be able to make a deal with the Wheat Pool that their agents would take delivery of the wheat on our behalf and store it in our bins.

Altogether this gave us a very large capacity, and our farmers were very cooperative and promptly filled it to the ceiling. Wheat at the time had a nominal value at the elevator of seventy-two cents, and in my discussions with the farmers I did all my mental arithmetic on that basis, hoping that in a year or two or three I would realize a price of seventy-two cents.

I have said there was no one at the office who had the slightest experience. My old, reliable secretary, Miss Allonby, had retired from my services on account of old age and had moved to Victoria. Her place was taken by two young ladies, Miss Olive Keating, who later married my son, Kevin, and was now his widow, and Miss Doreen Bushell, the daughter of a Limerick farmer. They kept the wheat accounts and did an outstanding job of accounting in a field where they were entirely green.

Nonetheless, by working long hours and figuring as hard as we could, we got a perfect record of the wheat going into the bins. Our friends at the

Wheat Pool came to our assistance nobly in that direction. Their agents received the wheat in the Pool elevator, weighed it, graded it, and issued our storage ticket to the farmer.

The Pool went further. Their agent accompanied the truckload of wheat to our storage bins and supervised the unloading of it therein. They then sent the storage tickets to Limerick periodically, and there the stenographers credited them to the various loans and kept track of them in every way. It was a very threadbare and skeleton setup, and the marvel is that it worked not only well but to perfection.

We had a great many worries in connection with this wheat. We had anticipated that we might have to carry that wheat for three or four years, and keeping it in condition and guarding it against fire were both live subjects for us. Even in our hands, some of it was not in very sound storage. Some of the buildings were not too weatherproof and required constant inspection. Again the Wheat Pool came to our assistance. They had a superintendent at Assiniboia who made a monthly tour of his territory, because they had all kinds of ramshackle storage too. While he was probing and testing their storage, he did the same for our bins, with the result that every month we got a report from the Wheat Pool that our wheat was in good shape. This was a service which was very much appreciated at our end.

We had altogether in storage scattered all over the country perhaps 400,000 bushels, a very important item in the picture of what was then a small company, Victoria Trust and Savings Company.

Victoria Trust was very pleased with the Saskatchewan situation. At the 1943 annual meeting, held in Lindsay on January 14, 1944, President T.H. Stinson reported that

> Our Company is very fortunate in having a very capable Western Manager in Mr. Wilson. He is able; energetic; a diplomat; possesses a pleasing personality and, while a good collector, knows how to keep the good-will of our borrowers. In 1943 we received in cash collections from Saskatchewan $180,000 and we have $200,000 worth of grain in our possession in storage for sale. This is a fine showing as it represents a return in one year of more than one-fifth of our total investment in that Province.[3]

In the early part of June 1944, I received a letter from the Canadian Wheat Board telling me I could dispose of this wheat provided I could get it all into the elevators by the 31st day of July 1944.

This presented a bit of a problem too. The wheat had come from I do not know how many different farmers and was supposed to be sold in the permit book held by the man who grew the wheat. The mere assembly of all these numerous permit books represented a lot of work, and I took that matter up with the Wheat Board. Their reply was to send me a general letter of authority to sell the wheat at any elevator convenient to us without paying any attention to the permit books—a privilege which saved us a lot of driving and was very much appreciated indeed.

We went out in the country and hired every idle truck we could find and started an army of men moving the wheat from our storage to the elevators, any elevator in which we could find space, and so diligently did we work that on the 31st day of July we did not have a bushel of wheat on hand.

In return for the courtesy and assistance we had received from the Wheat Pool, we gave them the business insofar as their facilities were able to take care of it. We must have been by far the largest customer the Wheat Pool had in 1944.

The Pool well understood that time was of the essence and that the matter was urgent, and they did not object at all, when they had no storage available, to the fact that we took the grain to their competitors. We just rushed it out regardless of everything.

When we disposed of all this wheat, there was no flaw in the accounting. Moreover, our weights held out. We did not lose a bushel that way, and even the grades which had been assigned to the wheat by the Pool elevator operators held out also.

Then, in July 1944, the cash settlements from the different elevator companies commenced to pour into the office. They were of the greatest joy to me, and I am sure they must have been to head office. We had used the figure of seventy-two cents a bushel, the amount of the initial payment which was being made at the local elevators, when we started the enterprise. But by the time we got around to delivering the wheat, this figure had been increased to $1.12 a bushel. The cash tickets were accompanied by a participating certificate, to which none of us attached much importance *at the time. But, a few months later, these participating certificates were called in, and we were paid an additional twelve cents a bushel.*

That made a price of $1.24 a bushel—of course, for Number One wheat at Fort William, and much of our wheat here was not Number One. Against the guide figure of seventy-two cents which had entered into all our calculations, you will have no trouble in appreciating that this was a most acceptable situation. I do not know the total of our remittances to head office in July 1944, but it must have been a very large amount indeed.

Victoria Trust's annual report for 1944 states that "$490,000.00 were collected by our Saskatchewan Manager, Mr. Charles Wilson."[4]

This wild adventure in wheat turned out to be a most happy business, thanks a lot to a great many good friends all along the line and certainly to the Saskatchewan Wheat Pool. The Victoria Board of Directors should be congratulated on the fact that they displayed vision and were willing to enter upon this adventure. Not only did they enter upon it, but they made it a real pleasure to work with them. I was given an absolutely free hand at this end, which was essential. If I were at Wood Mountain and a problem were presented to me, I had to make an immediate decision. There was no time to submit the problem to the board for their decision. I cannot recall one single occasion on which I received a letter criticizing what we were doing or giving us anything but the warmest support. And the good God knows we needed that, because those were tense and anxious days, and we had the feeling here that we were abroad upon quite rough and uncharted seas. That the adventure ended so happily and fortunately has always been one of the supreme joys of my life.

ENDNOTES

1. Wilson, *A Century of Canadian Grain*, 736.
2. LAC, RG 19, vol. 427, file 105-1A-12, Report on the Progress of the Farmers' Creditors Arrangement Act to March 31, 1941, Schedule 2.
3. Victoria Trust and Savings Company, *Forty-Eighth Annual Report* (Lindsay, ON: Victoria Trust and Savings Company, 1943), 10.
4. Victoria Trust and Savings Company, *Annual Report* (Lindsay, ON: Victoria Trust and Savings Company, 1945), 13.

CHAPTER 22

THE WAR ENDS *and* PROSPERITY RETURNS

T he provincial election of June 15, 1944, elected a CCF government and ended the Liberal Party's rule of Saskatchewan.
Liberal governments had presided over the province since it
was formed in 1905, with the sole interruption of the Anderson Co-
operative government in 1929–34.

Led by T.C. (Tommy) Douglas, the CCF nearly annihilated the
Liberals, winning all but five of the fifty-two seats. Three Active Ser-
vice MLAS were also elected. For the first time ever, Charles' riding,
Notekeu–Willowbunch, was no longer represented by a Liberal. The
long-time Liberal MLA, Charles Johnson, of Meleval, the town next to
Limerick, was soundly defeated.

The election should have been held in 1943, the traditional five years
after the 1938 election, but the Patterson government had extended its
term for one year, using the excuse of wartime conditions. The decision
became an election issue and today is considered to have worked very
much to the advantage of the CCF, whose strength was burgeoning.

When Premier Douglas and his cabinet were sworn into office
on July 10, 1944, much was made of the fact that Saskatchewan was
then governed by the first socialist administration in North America.

Charles was naturally disappointed with the defeat of his beloved
Liberals but not surprised. Three years earlier, on October 21, 1941,

Liberal committee rooms in Limerick, 1944.
Photo in possession of the author.

he had predicted such a loss in a long letter to Minister of Agriculture James G. Gardiner: *The rural sections are and have for some time been full of discontent ... no one mainspring of dissatisfaction, but just one thing here, another there but all adding up to a sullenness, a discontent, a sad lack of any of the old time Liberal enthusiasm. ... This* "discontent," Charles had reported, was being capitalized on by the CCF at meetings throughout rural Saskatchewan. *Believe it or not, after over two years of war, the rural people are only remotely aware of the fact that we are at war.* He had theorized that the casualty lists had not yet begun.

Charles had recommended that federal representatives counter with a series of meetings designed to emphasize that many of the unpopular national policies were dictated by the necessity of *a very real and terrible war. ... Set out your policies so that they must be seen and judged against the background of war. You will take another long step in building a nation by uniting every rural home in common sacrifice in defence of our common country.*

And, as a by-product or consolation prize, you can possibly salvage the government of Saskatchewan, whose present moment chances of surviving an election are practically invisible.[1]

To underscore the fact that a new order had arrived and that the office of Wilson Brothers no longer enjoyed easy access to, much less influence with, the provincial administration, the new CCF government cancelled Charles' appointment as a notary public.

But in Lindsay, Ontario, the directors and officers of Victoria Trust looked on Charles as the Golden Boy. He had pulled the company's chestnuts out of the fire of the Dirty Thirties, and the unusually large remittances of 1943 and 1944 were well appreciated. At the forty-ninth annual neeting, held in Lindsay on January 11, 1945, every speaker, the president, the general manager, and both vice-presidents, lauded their western manager.

President T.H. Stinson, KC, advised that "In the years 1943 and 1944 we have collected from our Western borrowers $800,000 which is slightly over 40% of our total Western investments." He included a portion of Charles' reporting letter: "Practically every individual loan we have left has been reduced to a figure where neither the Board of Review, the Provincial Mediation Board, nor the Provincial Government can cause us much worry in the future."

C.E. Weeks, the general manager, declared that

> Our Saskatchewan mortgages and agreements have been so greatly reduced individually and collectively in the last three or four years and the morale and resourcefulness of the western farmer is now so high through success won largely by his own pluck and perseverance that except where inherent weakness may exist or develop we have no more reason to feel anxious about our Western mortgages than those in Ontario, or to have any feelings of "inferiority complex" towards them.

J.B. Begg, the first vice-president, was effusive:

> Now there is one gentleman to whom I must refer and no doubt most of you have seen him at some of our past meetings. I refer to Charlie Wilson, our agent in Saskatchewan, and while you may have seen him very few of you know him. There is just one "Charlie Wilson" and he has more angles to his way of doing business for the Company than anyone could imagine. He appears to know each individual borrower and knows how to approach him to get the best results for the Company, and not

only that, but he is a personal friend of each and every one of them, and they will all swear by "Irish Charlie Wilson."

From H.J. McLaughlin, KC, the second vice-president: "Particularly insofar as the excellent results from Saskatchewan are concerned our thanks and appreciation should go to the industry of those two men (the President and General Manager) and to Charlie Wilson, the most able, persuasive, good-humoured and effective loan manager that any investment Company ever had on the Canadian prairies."[2]

Charles told the directors that the credit they were bestowing on him actually belonged to his son, Kevin, who had initiated the concept of farming the company's vacant lands.

The policy [of accepting wheat on account of mortgages] *was really an extension of the practice which Kevin started here. This episode was the highlight of my business career in the South Country but one for which Kevin was much more entitled to the credit than I was. I simply widened the trail which he had already blazed.*

The return of decent rains, but also better farming methods, brought about prairie agriculture's successful escape from the Dirty Thirties. Montana's Agricultural Museum in Fort Benton credits three Canadian inventions as rescuing dryland farming: summerfallowing, to conserve moisture; strip farming, to combat wind erosion; and the Noble plow, which killed weeds at the root level but left them as a protective trash cover on the surface.

In 1943, the publication of *Plowman's Folly*, a treatise debunking the custom of deep soil cultivation that had prevailed for years, revolutionized farmers' thinking. "The truth is that no one has ever advanced a scientific reason for plowing"[3] at first was considered near heretical but soon worked a huge impact on farming practice. A different method, calling for as little cultivation as possible, with all plant growth returned to and absorbed through the surface of the soil to provide needed nutrients, became the ideal. Charles became an early adherent of the new philosophy and kept extra copies of *Plowman's Folly* in the Wilson Brothers office as he attempted to convert others.

Not every year after the Dirty Thirties passed enjoyed adequate rainfall, but with improved methods farmers were better able to cope with dry conditions. The occasional year of low precipitation still occurred. In 1946 and 1947, some areas in the extreme southwest received even less rain than in 1937.[4] In 1961, Assiniboia recorded 67.8 millimetres, or 1.2 inches, of rain in May and only 17.3 millimetres, or 0.68 of an inch, in June and 12.4 millimetres, or 0.48 of an inch, in July. But soil erosion was now under control, and the dust storms of the 1930s did not return.

In later years, farming methodology evolved to the point where zero till became the mantra, and even summerfallowing and strip farming became passé. With the advent of air seeders, which insert the seed with a minimum of surface disturbance, and the use of herbicides and fertilizers, continuous cropping is now possible, and soil erosion has become an event of the past.

In 1945, a federal election was expected. Dr. Thomas Donnelly, Charles' friend and the long-serving MP for the federal constituency then called Wood Mountain, would not be running. After twenty years as an MP, he would accept an appointment to the Canadian Farm Loan Board.

Charles was under strong pressure from the local Liberal executive to stand as a candidate to succeed Donnelly, but he resisted, claiming that he was needed at Limerick to attend to the affairs of Wilson Brothers and Victoria Trust and Savings Company. When Charles Johnson was defeated as an MLA, Charles immediately promoted him as the logical federal Liberal candidate.

And so it was. Johnson became the Liberal candidate for Wood Mountain, only to be defeated again in the election of June 11, 1945. The CCF now represented the area both in Regina and in Ottawa. Its new member of Parliament, Hazen Argue, would become a very controversial politician. Although he and Charles were early opponents, in 1962 they became political bedfellows when Argue, then still an MP, left the CCF and joined the Liberal Party.

Prime Minister Mackenzie King was re-elected with a slim majority with 118 seats, to sixty-six seats for the Progressive Conservatives, twenty-eight for the CCF, and thirteen for the Social Credit, but his

Liberals did not do well in Saskatchewan, and King lost his own seat in Prince Albert. Saskatchewan elected only two Liberal MPs as the CCF took eighteen of the twenty-one ridings, leaving John Diefenbaker the only Progressive Conservative elected.

As well as it did, the CCF was sorely disappointed. Polls before the election had predicted that the party would win seventy to 100 ridings across Canada and perhaps even enough to form a minority government.

One of the new CCF MPs was Ross Thatcher in Moose Jaw. His father, Bill Thatcher, had owned a hardware store in Limerick from 1918 to 1928, and the family had lived next door to the Wilson home. Ross and Kevin Wilson, both of an age, had been friends, had raced their kiddie cars in the Wilson veranda, and had been active Young Liberals together. Now Ross was another political opponent. But in 1956, he left the CCF, rejoined the Liberal Party, and, when he contested the Assiniboia riding against Argue in the 1957 and 1958 federal elections, was welcomed back by Charles.

The war in Europe ended with the German surrender on May 8, 1945, and in the Pacific with the Japanese surrender on August 15th. At Limerick, Charles organized a committee to provide welcoming ceremonies for the community's returning service men and women. From the small village and the surrounding RM of Stonehenge, 293 men and women had served in the Canadian Army, the Royal Canadian Navy, and the Royal Canadian Air Force. Twenty-two had not returned.

Not content with the concept of a mundane cenotaph, the community determined to create a living memorial. A two-storey residence was acquired, renovated, and equipped, and on August 23, 1946, Lieutenant Governor R.J.M. Parker officiated at the dedication of the Limerick Memorial Hospital. As did others, Charles and Florence donated a ward in honour of Kevin.

In the summer of 1945, Charles drove out to Stonehenge to visit Ted Oancia and the farm that he had rented for some fifteen years. Son Garrett, then just thirteen, went along for the ride. When returning to Limerick later in the day, Charles mentioned that Ted wanted to purchase the farm, disappointing news because he did not want to sell the land.

Garrett naively asked, "Well, if you don't want to sell the farm, why do you?"

"It is their home. They are entitled to have it" was the firm reply.

It was the response of an Irish peasant whose people had served for generations as tenants on land that they could never own. Only a little more than twenty years before that day in 1945 had Irish Land Reform enabled Charles' brother William to purchase Driem, their family home for generations.

Ted and Fannie Oancia did purchase the farm, which in time passed to their son, Clarence. Oancia Farms is today a modern and prosperous agricultural enterprise.

By 1945, the Depression was over, but it left loan companies owning a great deal of Saskatchewan farmland. Victoria Trust and Savings Company was not excepted.

I went back to being a field man, looking after the mortgages and trying to dispose of the real estate, of which we had too great an accumulation, at least in the opinion of the Ontario inspector of trust companies. We made many deals in which we traded land for wheat.

My good friend Art Labay, of Killdeer, had a well-established farm on the south half of Section 9-2-3 W3rd. We owned the north half of that section, or at least 280 acres of it, having sold forty acres to the CPR for the townsite of Killdeer. I met Art in Wood Mountain one day, and he remarked, "I hear you are trading land for wheat."

I told him, "In the odd case, yes."

"How many bushels of wheat do you want for that north half of mine that I'm renting from you?"

I had hopes of realizing $2,500 out of that half section. Wheat was selling for seventy-two cents in the elevator, when it could be sold. I did a little swift mental arithmetic and said, "I want 5,000 bushels."

Art's reply was in the vernacular. "You can go to hell. I will never give you that. When are you coming to Killdeer?"

"I hope to be in Killdeer at noon next Tuesday," I replied.

"Come over to the house and have lunch with us when you get there. Maybe you and I might still make a deal."

Well, the following Tuesday I was in Killdeer and was glad to go into that lovely farm home for lunch. Art had three combines working in the fields, with three crews, and they fed those men first and turned them out in the yard, and then Mrs. Labay and Art and I sat down for a leisurely lunch, and a lovely lunch, as I recall still.

Knowing Art to be a very shrewd horse trader, I did not bring up the subject of the land, but about the time we got around to pie and coffee Art did. He said, "I will give you 3,500 bushels of Number Two wheat for that land."

I did not hesitate. I looked him in the eye and said, "Art, you have just bought yourself a farm."

His reply was "Excuse me for a moment, please." He went to the door and hollered at his crews, who were still in the yard. "Don't bring any more wheat over to this yard. Take it to the Victoria Trust annex." Then he returned to complete his lunch.

By the time we were through enjoying lunch, the first of the wheat was pouring into the annex at Killdeer, and by the following evening the whole 3,500 bushels had been delivered, and I had on my hands the duplicate of the tickets for it.

I requisitioned from head office the title and the transfer and delivered them to Art, and that was the end of the matter as far as I was concerned. However, some time later I got a letter from the income tax people asking me if I would be kind enough to allow them to inspect the agreement under which Art Labay had purchased that land.

I had forgotten there was no agreement, and I spent an hour searching for a non-existent agreement until my memory told me I was wasting my time, whereupon I wrote the income tax people and told them, "In this instance, we regret that we are unable to meet with your wishes, because no agreement of any kind, other than verbal, ever existed."

Charles' diary contains some comments on the terrible winter of 1947 and its impact on CPR train service. Normal operation on the Assiniboia–Shaunavon line was a train from Moose Jaw to Shaunavon six days a week, returning the following day. Thus, Limerick had two trains a day, one westbound, one eastbound.

Tues. Jan. 14: Bad day—ten to twenty below and blizzard. Train did not leave Moose Jaw; other train arrived and turned around at Assiniboia.

Thurs. Jan. 16: Weather has moderated and is now around zero or better with no wind. The Moose Jaw train arrived at 2:30 a.m. this morning, after having missed a day entirely. Set off no mail. We have had no papers and no mail since last Monday. First time since Limerick was born. Highways are blocked and railway snow fences covered in some places.

Fri. Jan. 31: A bad blizzard and thirty below. A really tough day and very hard on coal.

Mon. Feb. 3: Sunday morning was fair and not cold. About noon a blizzard came up from west and turned into a very bad storm. It lasted well into Monday morning, and at noon Monday there is still some of it, and it is about twenty below with NW wind. A mean day. Regina had a bad one, and stores are not open there today. No trains.

Tues. Feb. 4: Last night was twenty to thirty below with no wind. Quite OK. Today is warmer, clear, and bright. Promising warmer. No trains.

Wed. Feb. 5: Overcast and mild. The night was mild and lovely. No trains.

Thurs. Feb. 6: A cold morning, and a bad storm came up later. I did not come to the office in afternoon. Too wild. No trains.

Fri. Feb. 7: Storm blew out in night, but today is cold with quite a wind, NW. No trains.

Mon. Feb. 10: Sunday was a very nice day, zero or above. Today is calm and clear and not cold. No trains.

Tues. Feb. 11: Calm, bright day but cold. Train Shaunavon to Assiniboia and back.

Fri. Feb. 14: Very mild day, but line to Moose Jaw not open yet. Mail and passengers are taken via Meyronne and Swift Current.

Sat. Feb. 15: This was a soft day and thawed quite a bit.

Tues. Feb. 18: Reasonably cold day, no wind. Trains returning to normal.

The two weeks without trains reduced Limerick's food supplies to a desperate level. Even yeast for bread making became precious.

ENDNOTES

1. Charles Wilson to J.G. Gardiner, October 21, 1941, in possession of the author.
2. Victoria Trust and Savings Company, *Forty-Ninth Annual Report* (Lindsay, ON: Victoria Trust and Savings Company, 1945), 7, 13, 16, 18.
3. Edward H. Faulkner, *Plowman's Folly* (Norman: University of Oklahoma Press, 1943), 1.
4. McManus, *Happyland*, 220.

END NOTE

In late January, 1949, Charles and Florence entrained for Toronto. They were invited to attend the Victoria Trust & Savings Company Fifty-third annual meeting in Lindsay on February 1, 1949, something Charles had done several times in the past, but this year he had been requested to address the meeting, a signal honour. The Wilsons were entertained at Lindsay for several days following the annual meeting, with dinners at the homes of C.E. Weeks, KC, then Chairman of the Board, and T.H. Stinson, KC, President.

Friction had developed between the CCF government and the Dominion Loan and Mortgage Association over what the Association viewed as restrictive legislation. As a result, almost all farm loan firms withdrew from Saskatchewan, Victoria Trust among them.

At the annual meeting, Charles spoke to the shareholders of the quality of "decency," which, he said, characterized the Company:

> "This quality of decency has always been expressed through your Board of Directors to the people in my province who were the debtors of your Company. Not only do I express appreciation for that fact, but I hope as I go along, to be able to indicate to you that this policy paid rich dividends to the Company in calling out a similar quality of decency on the part of all its borrowers...

"I say to the shareholders of this Company that if they wish to return to my province and extend their business again, the way is wide open for them to do so, and they will have an excellent reputation to start with, and every borrower who did business with this Company before, would be prepared to do business with them again. It may be that this good-will will never be taken advantage of, but good-will is always an asset... ."[1]

Very few Ontario loan companies possessed any measurable good-will among Saskatchewan farmers after the disastrous experience of the 1930s. Many, perhaps most, Saskatchewan farmers held their lenders in very low regard, viewing them to be responsible for much of the general misery that had been suffered. That this was often unfair mattered not at all. It was a variation of the "Goddam the CPR" syndrome.

Victoria Trust made no attempt to capitalize on its good-will in Saskatchewan. In 1950 it merged with The Grey & Bruce Trust & Savings Company to form Victoria and Grey Trust Company. More than thirty years later, in the early 1980s, the new company did return to Saskatchewan, long after the good-will Charles had spoken of had been lost, but when it judged the political and business climate to be more propitious. In its publicity announcing its entry to Saskatchewan, no mention was made of the fact that Victoria Trust had a lengthy history in the province. In 1984 the Company merged again, this time with National Trust, and then was swallowed up by Scotiabank.

ENDNOTES

1. *Fifty-third Annual Report*, Victoria Trust & Savings Company, February 1, 1949: 27, 29.

EPILOGUE

In 1946, Charles was diagnosed to have a tumour on his spine and advised by his medical specialist, Dr. David Roger, of Regina, that the world-famous Dr. Wilder Penfield, at the Montreal Neurological Institute, was the only surgeon capable of correcting the problem, if it could be corrected.

So, in November, Charles put his affairs in order, and Florence drove him to Regina, where he climbed aboard his first airplane, a TransCanada North Star. The plane stopped briefly in Winnipeg, where Charles visited with his daughter Sheila, who was studying interior decorating at the University of Manitoba.

The operation was a complete success, and Charles was home for Christmas.

The following year Florence suffered a heart attack, not major but requiring hospitalization and rest before making a complete recovery. Both Charles and Florence enjoyed good health thereafter for many years, quite a few years in Florence's case.

In the 1950s, Charles and Florence rediscovered Victoria and began to spend a month there, or sometimes more, in the late winter. But they did not return to the Empress Hotel. Instead, they booked into a modest apartment hotel on Douglas Street.

They maintained their connections with a large circle of friends, many of whom they found similarly enjoying the winterless climate of Victoria. That led to bridge sessions, dinners, and just visiting.

Charles and Florence in Victoria, circa 1950. Photo in possession of the author.

Charles' passion for gardening took them often to walks through Beacon Hill Park and the Butchart Gardens.

Many of their friends had, like Charles, once been Saskatchewan homesteaders, and they considered themselves members of an unofficial but elite fraternity, wearing their experience proudly, like a badge of honour. A few had succeeded on the land, most had moved on to other occupations, but all had overcome many obstacles, including the Dirty Thirties, and had achieved a measure of prosperity. Charles teased W.C. Wells, who had created a successful construction company. *Billy Wells is the only homesteader to become a millionaire.*

In the spring of 1954, Charles returned to Ireland, half a century after he had last seen it. The incentive for the trip was an invitation to attend the dedication of a memorial at Floriana, Malta, commemorating the nearly 2,300 Commonwealth airmen who lost their lives in the Mediterranean Theatre in the Second World War and had no known graves. Some 285 members of the Royal Canadian Air Force were recorded, including Flying Officer Kevin Davies Wilson.

Charles and Florence first flew to London and then to Dublin, and then they travelled south to County Wicklow, Rathdangan, and Kiltegan, and they paid a visit to Driem, Charles' family home. Charles found his return to be very emotional. *It was as if I never left.*

They returned to London in time to be flown by the Royal Air Force to Malta for the dedication of the memorial on May 3, 1954, by Queen Elizabeth II, then just two years on the throne. The Malta Memorial is a fifteen-metre marble column topped by a gilded bronze golden eagle. At the base of the column, a brass plate bears this inscription: "Over

Charles in a wheat field, 1965. Photo in possession of the author.

these and neighbouring lands and seas the airmen whose names are recorded here fell in raid or sortie and have no known grave."

In the 1950s, Saskatchewan initiated a program of naming northern lakes, rivers, and other geographical features after the more than 4,000 Saskatchewan service men and women who lost their lives in the Second World War. Wilson Lake, approximately 150 miles north of Lac La Ronge, was dedicated to Kevin. When advised of the memorial, Charles requested that the name be Kevin's Lake, rather than the more common Wilson, but was unable to secure the change.

On August 25, 2001, Garrett Wilson and his son Kevin, named after his lost uncle, flew into Wilson Lake with Doug Chisholm of Woodland Aerial Photography in La Ronge, who has assisted some 225 families in commemorating these geomemorials. Garrett and Kevin fished the lake, camped overnight, and to a prominent rock outcrop affixed a brass plaque commemorating the dedication of Wilson Lake to Kevin Davies Wilson.

Charles and Florence returned to Ireland in 1966 to celebrate their fiftieth wedding anniversary. On the morning of August 1st, they were surprised in their room in the Shelbourne Hotel in Dublin to receive a then rare conference call from their three children back in Canada, Moira then in Victoria, Garrett in Regina, and Sheila in Sudbury.

Charles passed away at Limerick on April 14, 1970, at the age of eighty-six. Florence continued to reside in the family home at Limerick and with some assistance did an excellent job of maintaining the large garden. She remained in the home to which she had come as a bride for seventy-two years until infirmities forced a move to Regina. There she celebrated her 100th birthday at a large gathering organized by her surviving children, Sheila and Garrett. Florence died on April 2, 2005, just six weeks shy of her 102nd birthday.

In 1945, Moira Wilson married a British serviceman, Arthur Vine, whom she had met in Washington during the war. They moved to Johannesburg, South Africa, where they resided until 1954, when they returned to Canada, to Burns Lake in northern British Columbia. Moira died at Burns Lake in 1984.

In 1949, Sheila Wilson also married a serviceman, Robert McMullan, an Irishman who had served in Egypt with the Royal Air Force. They resided in Sudbury and Manitoulin Island, where Sheila, now widowed, lives alone and still operates the seasonal art gallery that she and Robert established more than thirty years ago.

Garrett Wilson, a retired lawyer who has turned to writing, lives in Regina.

Thomas Wilson worked for some years as a loan inspector with Canada Life after he left Limerick in 1923 and then established a swine-breeding farm near Pleasantdale, Saskatchewan. In the 1950s, he retired to Vancouver, where he died on August 12, 1965.

Ben Lloyd, who journeyed to Canada with Charles in 1905, homesteaded at Viscount, Saskatchewan, where he established a successful farm implement agency. He then engaged in land speculation, with disastrous results. In the late 1930s, he briefly attended the University of Saskatchewan. By 1951, he was living in penurious circumstances in Saskatoon, but his whereabouts thereafter are unknown.

THE LIMERICK PLAN

The problem of making arrangements to meet the situation which will arise in the Fall of 1932, if we are fortunate enough to harvest a reasonably good crop and the price of wheat remains at its present low level, is one to which the Board of Trade in the Village of Limerick, in collaboration with the farmers of the surrounding district, has given considerable thought and study during the present winter. From this study there has emerged a plan which we believe will fairly well meet the situation. To give credit where credit is due, it is only fair to state that the broad outlines of the plan were first crafted by Mr. H.E. Drope of Regina, and the details, insofar as they have been filled in, have been filled in at a couple of meetings which have been held in the Village of Limerick.

Before going into the details of the plan, it would perhaps be well to outline our view of the problem which we are attempting to find a solution for. Our district has suffered from three crop failures, at least the crop of 1929 was a very short crop, the crop of 1930 still shorter, while 1931 amounted to almost nothing. In 1928 our farmers delivered to the local elevators around eight hundred and fifty thousand bushels of wheat, while in 1931 this fell to four thousand bushels. Farming in our district has been carried on at a loss during those three years. In the first portion the loss was largely taken up by additional advances from the Loan Companies and Banks. When these became no longer available, the Municipalities had to step into the breach, while, later on, the Government had to take over from the Municipalities. The capital which has been advanced in this way during the past three years represents a debt of large proportions which would not have been incurred in the ordinary course of agricultural experience. At the same time, interest on the debt existing prior to the crop failures has gone into arrears, while most of the taxes have also gone unpaid.

There have been few purchases of new machinery during the years 1929, 1930 and 1931, and during the last two years only the most essential repairs have been within reach of the farmer. Due to these causes, the physical equipment of the farms is, at the moment, in an unusually run-down condition, and no small amount of capital is required to put it back into good working shape. Added to that is the fact that there has been little if any replenishing of the necessities of the household in the shape of boots, shoes, clothing and other similar lines. Apart from the creditors already mentioned, that is the Government, the Municipality, the Loan Companies and the Banks, most farmers have obligations also to the Machine Companies, Oil Companies, Lumber Companies, and a scattered host of miscellaneous creditors. Not all of these are interested on every farm, but some of them are. Again, there is the further fact of which we think note should be taken. The farmers' nervous system has not escaped the wear and tear of the past three years in any greater degree than the systems of people in other walks of life.

In 1929, when there was little crop in our district, the opening of the collection season witnessed the putting into commission of a number of trucks by the small and unsecured creditors. These people camped at the farmer's place and worried him for at least one load of grain to apply on their debt, in most cases agreeing to haul it free of charge. This traffic attained quite considerable proportions and an accounting to the secured creditors of the crop which had been threshed. At the present time, the number of preferred creditors is such we foresee that the farmer will be "highjacked" out of a large portion of his crop by the same methods, and this is certain to give rise to irritation on the part of the secured creditors and misunderstanding between them and the farmer. Not only that, but the farmer will be harassed to death by interviews with a legion of collectors and insistent demands for bills of sale and chattel mortgages on all or a portion of his crop.

We have mentioned the need for replacement in the physical equipment of the farm and of the household. We might now point out that, in the years prior to 1930, it was customary for the farmer to have disposed of pretty well all of his crop by the end of the year. After January 1st, he went back to his Bank and borrowed money to finance the succeeding year. The indications are that in future years this source of supply will not be available, at least to many farmers, and it is therefore of importance that when the crop is harvested a sufficient reserve shall be retained on the farm to carry the business through the following year. In other lines of business it is considered good policy occasionally to plow a large portion of the receipts back into the business. At the present time, it would appear imperative that the same procedure shall be followed to some extent in the agricultural industry. That is the problem as we have outlined it to ourselves at Limerick, or rather as the farmers have outlined it to us. Obviously some form of organization is urgently required to reconcile the claims of competing creditors. We assume that the Relief Commission will discontinue operations when a reasonably good crop is in sight. It will therefore be necessary to have some organization to ensure the retention on the farm of the necessary supplies to carry on for a year. Such portion of the crop as is available to be applied in reduction of debt should also be distributed with a due regard for precedence in the claims of the various creditors. Lastly, it is worthwhile taking some steps to ensure that the farmers will not be harassed to the point of desperation and subjected daily to humiliation by numerous and insistent demands for security on his crop as soon as it is severed from the land.

The organization which we propose to set up to meet this problem is of a simple nature. If we have a hundred farmers in our community who have a sufficiently large number of creditors to bring them within the class we have been describing, we suggest that they enter into an unregistered partnership with each other, in an organization, which for want of a better name at the moment might be called the Limerick Bureau of Farm Management. The Bureau would elect a president, vice-president and secretary, in addition to a Committee of four or five. They would then proceed to rent an office and engage the services of a paid man of the calibre of the Municipal Secretary or local Bank Manager. Some little office furniture would be necessary and also the services of a stenographer. As we vision the matter, if at July 1st the prospects warrant us in expecting any kind of a reasonable crop, the office would then be put into operation. During July and August the manager would have plenty of work in procuring abstracts, G.R.

Certificates and statements to enable him to make an exact record of each item of the total debt of each of his subscribers. When the farmer has completed threshing his crop and delivering same to the elevator, he will sell the total amount in one lump and take the proceeds to the office. They will then figure out how the money should be divided between the different creditors, after first setting aside the amount required to carry on with during the next year and paying the harvesting expenses. If there are twelve creditors, the manager can then have twelve copies of the statement made out, the statement giving particulars of each debt, its nature and the security behind it and showing the amount which it is proposed to pay on it, the total amount of the crop, and the amount which has been set aside to carry on the business. One copy of this statement will be mailed to each creditor with a brief circular letter setting out the fact that, if no objections are received in ten days to the proposed method of distribution, cheques to carry such distribution into effect will be issued at that time. By this method every creditor would know the exact circumstances surrounding his debtor and exactly what became of the crop proceeds. If the creditor received nothing on account, the statement will assure him at least that that is the share which was coming to him. At the end of the year, the manager of the office would be able to hand to each of his subscribers a statement showing him exactly where he stands.

The question of the cost of the scheme will no doubt be asked. We believe that a competent man, under present conditions, could be engaged for a salary of $1,500.00 a year, and with the assistance of a stenographer for six months, we believe he could handle a hundred files. A stenographer for that period would cost $350.00. Allow $640.00 to rent a small office, pay for a minimum amount of office annually. Estimating the average farm income in our community at $2,500.00, this would represent about one per cent, but the farmers declare that, if it cost no more than two percent, there need be no objection to it. At least we believe that the scheme would cost less than the scores of collectors who will, in the absence of some such scheme, be busily engaged in our territory, and we believe, in the long run, the cost of these collectors goes back to the farmer.

We would summarize the benefits of this scheme as follows:

1. It is a dignified and straightforward method of dealing with an intimate problem. The farmer's business need be disclosed to no one except the manager of the office and the farmer's creditors. With a little cooperation on the part of the creditors, which we are optimistic enough to believe would be forthcoming, the farmer would be relieved of all interviews at the hands of collectors, while all creditors would get an absolutely square deal without expending any money in collection costs on the claim.

2. If, at a future date, as may well happen, there arises, in one case or in many cases, a situation which calls for an actual reduction of debt, an office such as we have outlined will have accumulated data in regard to the income-producing possibilities of the farm which should be an invaluable and intelligent guide in solving the problem of readjustment. As far as we are aware, no such data is in existence today.

3. We believe that the existence of such an office would be a definite step forward in providing for better management on the farms, and we believe the future will imperatively demand better and more careful management.

4. It is a first step in the direction of the farmers attending to the matter of their own salvation as opposed to the trend of the last couple of years, which has been one of going to the Government with every problem. This scheme would require from the Government nothing but their benediction, although the right of free searches at the Land Titles Office and Sheriff's Office would of course be appreciated.

5. Finally, it will preserve the Farmer's self-respect and peace of mind and pay good dividends in that direction.

Respectfully submitted on behalf of the Board of Trade of the Village of Limerick.

BIOGRAPHIES

GEORGE F. EDWARDS

George Edwards was born on June 7, 1878, at Essex, England, then a suburb of London. At the age of 3 he emigrated with his family to Cornwall, Ontario. In 1906 he came west and homesteaded near Earl Grey. After four difficult years proving up his homestead, he decided the land was too stony, so he sold and moved to better land at Markinch, where he was more successful.

Edwards had a natural talent for public affairs and administration and in 1912 he was elected reeve of the Cupar Rural Municipality, a position he held for 10 years. In 1919 he became vice-president of the Saskatchewan Municipal Hail Association. He then became active with the Saskatchewan Grain Growers Association (SGGA) and the Progressive political movement. Soon he was vice-president of the SGGA and deeply involved with the formation of the Saskatchewan Wheat Pool

In 1924 Edwards was elected president of the SGGA and in 1929, president of the Canadian Council on Agriculture. When the two organizations amalgamated as the United Farmers of Canada, Saskatchewan Section, Edwards was the first chairman and remained active for many years

Concerned about the extent of farm debt, in 1919 Edwards made a presentation to the Saskatchewan Liberal government requesting the establishment of machinery to adjust, or reduce, the debt of farmers drowning in red ink, perhaps the first time such a suggestion was made. In 1935 Edwards was appointed a member of the Saskatchewan Debt Adjustment Board, a position he held until 1943. There he was instrumental in reducing the debt of thousands of farmers.

His eyesight failing, Edwards retired from farming in 1951 and moved to Vancouver. Soon his old passion for public affairs resurfaced and he was elected to four terms as president of the British Columbia Senior Citizens Association. He died in 1976. In 1984 he was nominated to the Saskatchewan Agricultural Hall of Fame.

ANDREW HOSIE

An Ontarian by birth, in 1903 at the age of 13 Andrew Hosie moved with his family west to Brandon. There he completed his education and joined Central Canadian Insurance Company, which seven years later transferred him to Regina. When war broke out in 1914, he enlisted with the 5th Battalion and was posted overseas where, in 1917, he received a battlefield commission as a lieutenant.

Upon return to Regina after the war, he co-founded Drope & Hosie, which soon became the city's pre-eminent real estate and insurance firm. But Hosie continued his military activities, serving with the militia and becoming the commanding officer of the 12th Canadian Machine Gun Battalion and of the Regina Rifle Regiment, attaining the rank of colonel.

While serving overseas with the 5th Battalion, Hosie had become acquainted with Major Murdoch MacPherson, also serving with the 5th. In 1929, MacPherson, then a lawyer practising in Regina, became Saskatchewan Attorney General in the Conservative/Co-operative government of Premier J.T.M. Anderson. When MacPherson found among his responsibilities the newly created Debt Adjustment Board, he recruited Hosie to serve as chairman.

As Chairman of the Debt Adjustment Board, Andrew Hosie struggled with the problems of the ever-increasing debt load caused by the failure of crops and prices in the 1930s, finally coming to the view that more powerful federal legislation was required to cope with the calamity. His appointment was terminated in 1934 when the Liberals under James Gardiner returned to power.

In June 1935, when the On-To-Ottawa Trek reached Regina, Hosie, as Commanding Officer, was distressed to receive orders to ready his 12th Machine Gun Battalion. Fortunately, further orders never came.

Andrew Hosie led a distinguished life in the Regina community, serving in executive positions with a large number of charitable and service organizations. He died on November 16, 1972.

CHARLES AVERY DUNNING

A 16-year-old iron worker with little formal education, Dunning emigrated to Canada in 1902 and soon after homesteaded near Beaverdale, west of Yorkton. Concerned about the treatment received by farmers, he attended the 1910 convention of the Saskatchewan Grain Growers Association at Prince Albert. Without funds for better accommodation, Dunning slept on the convention floor, but his natural speaking ability and passion caused him to be elected a district director. The next year he was vice-president. Then he became the first general manager of the newly-formed Saskatchewan Co-operative Elevator Company which in only four years became the largest grain handling facility in the world.

Invited to join the Saskatchewan Liberal government, Dunning was elected in 1916 and became Provincial Treasurer, a position he held for 10 years, even after becoming Premier in 1922.

Dunning continued to advance and in 1926 was recruited to the federal cabinet of Prime Minister Mackenzie King as Minister of Railways. In 1929 he became Canada's Minister of Finance but was defeated along with the King government in 1930 and turned to the business world.

When Mackenzie King became Prime Minister again in 1935, Dunning was induced to return as Minister of Finance, a portfolio he held until 1939 when failing health forced his retirement. Turning again to business, he served as president of Ogilvie Flour Mills, held a number of corporate directorships, and became Chancellor of Queen's University. He died in Montreal in 1958 at the age of 74.

JAMES G. GARDINER

Born in Ontario on November 30, 1883, Gardiner came west with a harvest excursion in 1901. Educated at Regina Normal School and Manitoba College, he served as a school teacher. When he became principal at Lemberg, he bought a farm that became his permanent home.

Elected as a Liberal to the Saskatchewan Legislature in 1914, he was appointed Minister of Highways in 1922 and succeeded to the Premiership in 1926. Following the defeat of his Liberal government in 1929, he served as Leader of the Liberal opposition until the election of 1934 returned him to the premier's office. He was then recruited to Ottawa by Prime Minister Mackenzie King and became Canada's Minister of Agriculture, a position he held for 22 years, presiding over the severe difficulties presented by the Depression and the Second World War. Defeated as a member of Parliament in 1958, he returned to the farm at Lemberg where he remained until his death on January 12, 1962.

THOMAS C. DAVIS

Thomas Davis, born in Prince Albert on September 6, 1889, grew up in the city that was already a Davis fiefdom. His father, also Thomas Davis, had served two terms as a Liberal member of Parliament and sat in the Senate from 1904 until his death in 1917.

The younger Davis graduated from Osgoode Hall Law School in 1909 and began the practice of law in Prince Albert. After serving as an alderman for two terms, and then as mayor from 1921 to 1924, he was elected to the Saskatchewan Legislature in 1925. The next year he became Minister of Municipal Affairs and then Attorney General in 1927. After the defeat of the Liberal Government in 1929, he served in Opposition until 1934 when the Liberals returned to office. He again became Attorney General until 1939 when he was appointed to the Court of Appeal where he remained only briefly, moving to Ottawa in 1940 to become Deputy Minister of War Services.

In 1942 Davis entered Canada's foreign service, becoming High Commissioner to Australia from 1942 to 1936, then ambassador to China, followed by Japan and West Germany. In 1957 he retired to Victoria where he died on January 21, 1960.

JOHN A. MAHARG

An Ontarian by birth in 1872, John Maharg came west at the age of 18 and settled near Moose Jaw, becoming a grain farmer and cattle breeder. An activist in the farm movements, he served as the first president of the

Saskatchewan Grain Growers Association and continued in that office from 1910 to 1923. He was also the first president of the Saskatchewan Co-operative Elevator Company from its beginning in 1911, and president of the Canadian Council on Agriculture from 1915 until 1917, when he was elected to Parliament from the constituency of Maple Creek as an Independent. He was recruited to the Saskatchewan government of Premier William Martin in May 1921, was elected by acclamation in Morse, and became Minister of Agriculture, resigning in December 1921 following a dispute with Premier Martin. Maharg remained in the Legislature as Leader of the Opposition until 1924 when he left politics and returned to farming but sat on the board of the newly formed Saskatchewan Wheat Pool. He died in November 1944. Maharg was elected to the Saskatchewan Agricultural Hall of Fame in 1977.

WILLIAM MARTIN

After graduation from Osgoode Hall Law School, the 27-year-old Martin moved to Regina to take up the practice of law. In 1908 he was elected to Parliament representing Regina City. Re-elected in 1911, he was induced to return to Saskatchewan to assume the premiership of the Liberal government, which had been wounded by scandal. Dealing decisively with the problem, Martin led his Liberal government to re-election in 1917. As Premier, Martin had successfully avoided confrontation with the agricultural movement by recruiting farm leaders such as Charles Dunning, who joined the government with Martin in 1916. In 1921 the Premier recruited to his cabinet John A. Maharg, another prominent farm spokesman, but a dispute with Maharg led to Martin's resignation in early 1922. He was appointed to the Saskatchewan Court of Appeal.

In 1941 Martin was appointed Chief Justice but shortly after became custodian of Enemy Alien Property during the Second World War and spent much of his time in Ottawa. After the War he continued on the Court until his retirement in 1961. He died at Regina on June 22, 1970.

Archer, John H., *Saskatchewan, A History*, Western Producer Prairie Books, 1980, Saskatoon, Saskatchewan.

Barry, Bill, *Geographic Names of Saskatchewan*, 1998, Peoples Places Publishing Ltd., Regina, Saskatchewan.

Berton, Pierre, *The Great Depression 1929–1939*, 1990, McClelland & Stewart, Toronto, Ontario.

Better Farming Conference, Report, July 6, 7 & 8, 1920, Saskatchewan Archives, R-261.

Booth, J.F., *Measures for the Relief And Rehabilitation of Agriculture in Canada*, Journal of Farm Economics. Vol. 17, no. 1, February 1935.

Brennan, James William, *A Political History of Saskatchewan*, 1905–1929, 1976 Ph. D. Thesis, University of Alberta.

Britnell, George, *The Wheat Economy*, 1939, University of Toronto.

Britnell, George, "Saskatchewan, 1930–1935," *The Canadian Journal of Economics and Political Science*, Vol. 2, no. 2, May, 1936.

Britnell, George, "The Saskatchewan Debt Adjustment Program," *The Canadian Journal of Economics and Political Science*, Vol. 3, no. 3 (August, 1937).

Easterbrook, W.T. and W.B.H., "Agricultural Debt Adjustment," *The Canadian Journal of Economics and Political Science*, Vol 2, no. 3 (August, 1936)

Edwards, George, *Memoirs Of George F. Edwards*, 1969, Saskatchewan Archives Board, A-3.

Egan, Timothy, *The Worst Hard Times*, Houghton Mifflin, New York, 2006.

The Encyclopedia of Saskatchewan, Canadian Plains Research Center, Regina, 2005.

Fairbairn, Garry Lawrence, *From Prairie Roots: The Remarkable Story of Saskatchewan Wheat Pool*, 1984, Western Producer Prairie Books, Saskatoon, Saskatchewan.

Faulkner, Edward H., *Plowman's Folly*, University of Oklahoma Press, 1943.

Gray, James H., *Men Against The Desert*, 1967, Western Producer Prairie Books, Saskatoon, Saskatchewan.

Gray, James H., *The Winter Years*, 1966, MacMillan of Canada, Toronto.

Holloway, Godfrey, *The Empress of Victoria*, 1968, Pacifica Productions, Victoria, British Columbia.

Isern, Thomas D., "Gopher Tales, A Study in Western Canadian Pest Control," *Agriculture History Review*, Vol. 36, no. 11, 1988.

Jones, David C., *Empire of Dust*, 1987, University of Alberta Press.

Jones, David C., Editor *We'll All Be Buried Down Here; The Prairie Dryland Disaster 1917–1926*, Alberta Historical Society, 1986.

Journals of the Legislative Assembly of Saskatchewan, Legislative Library, Regina.

King, Mackenzie, *Papers*, Library and Archives Canada, Ottawa, J-1107-4G

Marchildon, Gregory P., "The Prairie Farm Rehabilitation Administration: Climate Change and Federal—Provincial Relations During the Great Depression," *Canadian Historical Review*, Vol. 90, June, 2009.

Marchildon, Gregory P. and Black, Don, "Henry Black, the Conservative Party and the Politics of Relief," *Saskatchewan History*, Vol. 58, no. 1, Spring, 2006.

McManus, Curtis R., *Happyland, A History of the Dirty Thirties in Saskatchewan 1914–1937*, 2011, University of Calgary Press.

McRae, D.B. and Scott, R.M., *In The South Country*, 1934, Saskatoon Star-Phoenix, Saskatoon, Saskatchewan.

Peel, Bruce Baden, *R.M. #45, The Social History of a Rural Municipality*, Master of Arts Thesis, 1946, University of Saskatchewan.

Saskatchewan Legislature, Journals, Legislative Library, Regina.

Schmalz, Wayne, *On Air: Radio in Saskatchewan*, 1990, Coteau Books, Regina.

Silversides, Brock V., *Prairie Sentinels: The Story of the Canadian Grain Elevator*, Calgary, Fifth House, 1997.

Smith, David E., *Prairie Liberalism: The Liberal Party in Saskatchewan 1905–1971*, University of Toronto Press, 1975.

Victoria Trust and Savings Company, Annual Reports.

Waiser, Bill, *Saskatchewan: A New History*, 2005, Fifth House, Calgary, Alberta.

Ward, Norman, and Smith, David E., *Jimmy Gardiner, Relentless Liberal*, University of Toronto Press, 1990.

Wardhaugh, Robert A., *Mackenzie King and the Prairie West*, Toronto, University of Toronto Press, 2000.

Watkins, Ernest, *R.B. Bennett, A Biography*, 1963, Kingswood House, Toronto, Ontario.

Wilson, C.F., *A Century of Canadian Grain*, 1978, Western Producer Prairie Books, Saskatoon, Saskatchewan.